DIVIDENDS FROM SPACE

DIVIDENDS FROM SPACE

Introduction by Wernher Von Braun and S. Fred Singer

Frederick I. Ordway III
Carsbie C. Adams
Mitchell R. Sharpe

THOMAS Y. CROWELL COMPANY
New York • Established 1834

 Published simultaneously in Canada by Fitzhenry & Whiteside Limited, Toronto.

Designed by Nancy Dale Muldoon

Manufactured in the United States of America

L.C. Card 70–170997

ISBN 0-690-24134-8

2 3 4 5 6 7 8 9 10

CONTENTS

INTRODUCTION

Wernher Von Braun and
S. Fred Singer

THE DEVELOPER'S VIEW

DURING THE DECADE OF THE 1960S, THE United States embarked on a space program that covered a broad spectrum of activities, culminating with the Apollo 11 and 12 landings on the Moon. Now, in the 1970s, one of our major challenges is to demonstrate that space systems can be applied here on Earth to protect the quality of the environment, to address the many problems deriving from overpopulation, and to enhance generally the quality of life for all mankind.

Space systems already have been utilized to help improve weather forecasting, to undertake national and international communications on an entirely new scale, and to assist in navigation. Itos satellites operated by the National Oceanic and Atmospheric Administration, for example, provide weather data to the National Environmental Satellite Service for worldwide analyses and forecasts, and the Nimbus program has shown the feasibility of obtaining global atmospheric temperature soundings from satellites. Continued advancement will permit, by about 1980, the development of comprehensive weather models with which we can make accurate 14-day weather forecasts and begin experiments in large-scale weather modification and climate control. Communications satellites, meanwhile, have evolved quickly from the Syncom of 1963 to the Intelsat IV models of today, each of which can provide almost 10 times as many channels or circuits as the most advanced transatlantic cable—and at a fraction of the cable cost. In the field of navigation, the Department of Defense has developed the Transit satellite, now being used by U.S. Navy ships and submarines, and the Department of Transportation has plans for an air traffic control satellite system. The latter concept, already demonstrated by NASA's application technology satellite, may be applicable to ships as well as to commercial aircraft.

With the development of satellites and remote sensors, we now have at our fingertips the tools necessary to survey and manage Earth's natural resources. This is a challenge of the first magnitude. Using multispectral sensor systems, it will be possible to take inventories of crops and timberlands and to monitor their health; to prospect for oil, natural gas, and minerals; and to mount programs for developing oceanic resources.

Potential hydrological applications are especially significant. For example, it appears likely that we will soon be able to map the thickness of snowfields. Knowledge of how much water will result at the thaw will help us to conserve our water resources by managing reservoir levels more efficiently. It has been estimated that an improvement of even a fraction of a percentage point in the accuracy of available-water predictions in the United States alone would save tens of millions of dollars annually. Spacecraft observation techniques offer the best hope of collecting such data on the large scale.

The potential of a global program for managing Earth resources to match the pattern of supply with the pattern of demand is so enormous that there is no question as to what must be done. Mother Earth can feed us all—even twice as many as we are now—if only we will stop exploiting her resources and start managing them. Can we afford to create a global resources management system? I think, and this book amply demonstrates, that we cannot afford *not* to develop one. Fortunately the nations of the world are approaching this task constructively. The United Nations Committee on the Peaceful Uses of Outer Space has encouraged rapid progress in surveying of Earth resources, especially among developing nations. Cooperative programs worked out by the United States with the Brazilian, Mexican, and Indian governments represent examples of this genuine interest in projects that promise tangible benefits to all concerned.

Satellites also can perform invaluable services in the field of pollution detection and control. Pollution of all kinds, whether of the atmosphere, of inland waters, or of tidal estuaries, can be measured and mapped more effectively by satellite than by any other means. While we will continue to need aircraft for pollution detection, because of the higher resolution always available from lower altitudes, spacecraft provide better overall views and are inherently cheaper to operate. It costs less to get an airplane aloft, but the spacecraft stays in orbit for years without any fuel consumption. The Earth Resources Technology Satellite (ERTS) program is NASA's first major step in this direction. It will be followed by the Earth Resources Experiments Package (EREP) in the Skylab space station program. Since men can operate remote sensors not now compatible with automated spacecraft and can select particular targets for observation, their presence in orbit will provide data with the highest informational content.

The potential applications of satellites are practically endless. In the area of Earth physics, key measurements have been obtained during the past few years to permit accurate determination of features on our planet in relation to one another, and better understanding of how terrestrial gravity affects orbiting objects. Such measurements are invaluable in constructing dynamic models of our sphere which can be used to predict, and to prepare for, earthquakes, tidal waves, and volcanic eruptions. Accurate ocean height measurements, meanwhile, can be helpful in forecasting changes in climate. We are even learning how to apply satellite sensing techniques to assist fishermen by locating schools of fish.

When assessing the dividends of the space program, we must not forget the great technological base that was created to put satellites and men into orbit. In the not-too-distant future, a manufacturer of automobile batteries may be able to say of his product: "It is permanently sealed, will last for the life of the car, will not leak, will not require filling, and will never be dead on a cold morning." Such a claim will be entirely honest; for much of the technology needed to manufacture this battery now is on hand, created largely through the efforts of space scientists and engineers to satisfy the unusual power requirements of satellites and spacecraft. And the auto battery is but one example among many of the down-to-earth applications of space technology. Already, as this book clearly shows, the space program has produced dividends in transportation and commerce, health and medicine, food and agriculture, and industry and manufacture.

Recently the question of the relevance of such scientific research to man's many problems here on Earth has dominated many public hearings. Scientists themselves are worried about it, as evidenced by the number of articles in professional journals on the subject. DIVIDENDS FROM SPACE, as the first complete report on practical benefits obtained from this nation's massive investment in the space program, could hardly be more timely.

WERNHER VON BRAUN
Deputy Associate Administrator
National Aeronautics and Space Administration

THE USER'S VIEW

WHEN SPACE EXPLORATION WAS FIRST CONceived, the primary motives were adventure, scientific investigation, and national prestige. Only with the advent of satellites have we become aware of the fact that our expenditures are actually producing an economic return—a real dividend from space. This is welcome news, particularly in view of the heavy investment that has been made during the past decade and of the need for further expenditures if space research is to continue. Even more welcome is the news that the dividends are going to grow with time while the investment is leveling off—a happy situation indeed.

Some of these dividends have been apparent to the general public; some of them only to specialists; and some of them have not yet even been perceived. This volume performs a valuable and unique service as it explains the real returns from space for today and tomorrow. The spinoffs of space technology are numerous and important, and benefit us in our everyday life in the form of instrumentation, computers, data handling systems, medical instruments, and so on. Perhaps one of the greatest achievements has been the demonstration that equipment can be made absolutely reliable—thus setting a new standard for industry in developing products which will last far longer without attention and without expensive and time-consuming repairs. Another unheralded spinoff has been the development of a managerial group of people who are able to analyze and handle complicated problems involving the bringing together of new technologies, of armies of specialists, of immensely complex test programs—all for the accomplishment of a single goal.

So much for technology. It is the *product* of space exploration which is now coming to the foreground—the results from satellites circling the Earth. Every schoolchild now is aware of the use of satellites for long-distance communication. Years from now we will realize that space communication has indeed changed the picture of our globe in the linking together of continents, not only through television but also through navigation and traffic control. The view of the Earth afforded by space cameras is giving us vital information about global weather, allowing us to make accurate predictions in areas where predictions have never been possible before.

These dreams are now becoming reality. Still in the future, but well founded, are applications to the surveillance and exploitation of Earth resources, for mapping, forestry, agriculture, fisheries, and efficient employment of water resources. Still further away is the utilization of the space environment which surrounds the Earth to achieve a better life for the population on our globe.

It is probably too early to judge. But in a century studded with all kinds of scientific and technological achievements—electronic, nuclear, genetic, to name a few—surely our exploration away from the Earth and into outer space may rank as the most momentous of all.

S. Fred Singer
Deputy Assistant Secretary
U.S. Department of the Interior

1
NEW PRODUCTS FOR HOME AND INDUSTRY

ON 7 OCTOBER 1968, JOSEPH BARRIOS, OF Morgan Hill, California, was shot in the head during a robbery. Rushed to O'Connor Hospital in San Jose, he was operated on immediately to remove the bullet, which had shattered into fragments as it entered his brain. An X ray revealed, however, that one fragment remained in an inaccessible and very dangerous location. It was floating around between the third ventricle and the upper ventricles—the cavities holding the fluid that drains from the brain into the spinal canal. If left there, the fragment could possibly lodge in the narrow passage between the ventricles and prevent the draining of the fluid, resulting in a fatal buildup of pressure on the brain.

Dr. Phillip M. Lippe, a neurosurgeon working on Barrios, remembered reading once that centrifugal force had been used in performing a delicate eye operation. He contacted fellow doctors and physiologists at the nearby Ames Research Center of the National Aeronautics and Space Administration, in Mountain View. This center had a centrifuge that was used for medical research and for the training of astronauts.

Learning what had happened, two doctors at Ames, Seymour N. Stein and Ralph Pelligra, began to work on the problem. First they ran some tests using a bullet fragment embedded in gelatin, which has about the same consistency as the brain; they needed to know how the fragment would move under acceleration. Then they carefully adjusted the seat in the compartment in which Barrios was to be moved to the center, so that the vehicle's acceleration would cause the fragment to move backward into the brain, where the doctors knew it could do no harm.

When he arrived at Ames, Barrios was fitted out with special bioinstrumentation that is used to measure heart and respiration rates and make electrocardiograms of astronauts and test subjects in the centrifuge. Properly positioned, the patient was monitored by a TV camera as the centrifuge began spinning around. For 55 seconds he whirled at a force of 6G—an acceleration that in effect increased his weight by six times. Afterward, X rays revealed that the bullet fragment had become embedded in an area where it could do no further damage.

Gunshot victim Joseph Barrios is fitted with biosensors prior to being accelerated to six times the force of gravity in the centrifuge at the Ames Research Center, in a successful attempt to reposition a bullet fragment in his brain. (NASA)

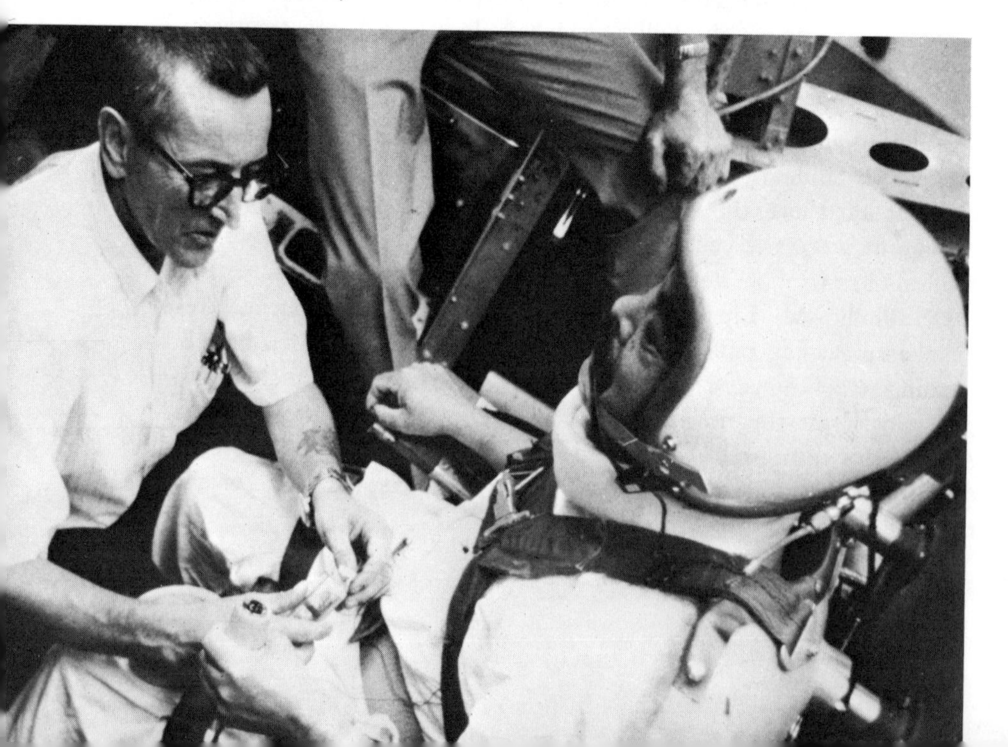

Joseph Barrios was the recipient of one of the more dramatic dividends resulting from aerospace research and technology. These dividends, often called spinoff or fallout, are found in a broad range of everyday human activities, from the home, business office, and factory to the doctor's office and hospital. Indeed, one of the least known aspects of the overall space program is the myriad ways that the man in the street benefits—all too often without realizing it. Improvements in medicine, communications, and weather forecasting; breakthroughs in man's ability to monitor and control pollution of his environment; new materials and more economical methods of production—these are among the areas in which significant advances affecting the quality of everyday life on Earth have been made, in the course of achieving such program goals in outer space as landing men on the Moon and taking close-up photographs of Mars.

Unfortunately, the National Aeronautics and Space Administration may have overpublicized manned spaceflight programs at the expense of its other operations, with the result that comparatively few of the citizens whose taxes support NASA's activities know very much about the practical benefits that have stemmed from them. The reasons for NASA's orientation are clear enough. The news media and the public at large generally are more interested in stories about people than in those about highly automated machines, and NASA naturally has catered to their interests.

Today reaction has set in, and NASA is paying a heavy penalty for its overpublicity of manned flight. Public interest in the space program, whetted by a series of progressively more exciting manned flights about the Earth in the Gemini and Mercury capsules, reached tremendous heights with the landing of Apollo 11 on the Moon on 20 July 1969. Thanks to the prior development of the communications satellite, the first Moon walk was witnessed on television by some 600 million people—by far the largest audience for any event in world history. But once the landing had been accomplished, public interest in the space program waned very quickly. The general attitude seemed best summed up in the idiom of show business: "That was great, but what can you do for an encore?"

Within a year of the Apollo 11 landing, NASA's budget was cut to its lowest level since 1963, and the entire space program came under intensified fire from a growing number of critics. While some people had maintained that the nation's priorities were askew practically from the time of President John F. Kennedy's address to Congress on 25 May 1961, in which he declared that the landing of a man on the Moon by the end of the decade ought to be a national goal, such criticisms seemed to have more point now that the drama of the event itself had faded. Had the thrill of seeing men walk on the Moon really been worth the $20-billion cost of the Apollo program? Were a few pounds of lunar rock and soil samples, together with photographs, seismographic readings, and other scientific data, worth such a steep price?

And the Apollo program, of course, represented only half the total funds spent on space research and exploration during the preceding decade. Some $40 billion, or 2.5 per cent of all federal expenditures during the 1960s, went for space. True, this amounted to just 0.5 per cent of the gross national product, but even such a proportion was said to be too much for a nation beset with a war abroad and poverty, unemployment, deteriorating cities, a fouled environment, and a host of other maladies at home. That Americans also spent huge sums on other items of questionable social value—$10 billion annually on hairdos and cosmetics, for instance, and $6.5 billion a year on alcohol—was dismissed as being beside the point.

Seeking to rejuvenate public enthusiasm for the space program, NASA followed the formula that had been so successful in the past and concentrated again on manned spaceflight. But times had changed. Talk of plans for manned space stations in orbit about the Earth, and for a new breed of reusable launching vehicles (called space shuttles) that would take off from conventional airfields and land again on them after delivering cargo and passengers to the stations, failed to evoke a sympathetic response from Congress. Citizens on the campuses, in the ghettos, and in the unemployment offices of the 1970–71 period had more immediate cares.

The space agency may have committed a serious strategic error in failing to give more attention to explaining the practical dividends accruing from the space program. In the long run, they would appear to make a far stronger justification for continuing the program on a large scale than all the new scientific data obtained from the instruments the astronauts left on the Moon and the rocks they brought back with them. The scientific data are of inestimable but theoretical value, and they are of interest mainly to scientists. The practical dividends, meanwhile, are already enriching the lives of everyone, and for this reason are more likely to win support for the program from Congress and the country.

FROM LABORATORY TO MARKETPLACE

Many products, materials, and new techniques developed for use in the space program have been converted into practical dividends in the form of new or improved products for home and industry, but it is often difficult to pinpoint the degree of space-inspired technology in such conversions.

There are several reasons for this. To begin with, the process of transferring technological advances from laboratory to marketplace is usually complex and time-consuming, and the aerospace industry is only one of a number of sources of new technological knowledge. The aerospace industry works in such close conjunction with other industries, independent laboratories, university research institutes, that the lines of distinction between individual contributions often are blurred. Moreover, progress in aerospace technology is but a concomitant of general technological progress, in which the aerospace industry is the beneficiary as well as the inspiration of advances in other fields.

Though at times generalized and difficult to describe with precision, the impact of the space program on other fields is nevertheless very real. A case in point is the inertial navigation system used in the Boeing 747 jumbo jet airliner. Although it is impossible to draw a clear line showing how equipment for airliners developed out of the Apollo hardware, the commercial Carousel IV system probably could not have been built but for the experience that the manufacturer, the Delco Electronics Division of General Motors Corporation, had previously gained in the space program. "We learned a lot about materials under temperatures and stress that opened the way for this commercial application," said Dr. J. H. Bell, director of reliability for Delco. "Apollo shoved us forward, made us aware of the need for new technologies; it accelerated all our thinking, our research and development techniques. As a result of that concentrated space effort, when the Boeing 747 autonavigator needs appeared, we were ready."

Sometimes the intangible dividends from space technology can be traced more easily to their source than in the case of the Carousel. For instance, after reading about NASA's work in solid state electronics, engineers at Hussman Refrigeration Company, St. Louis, began to wonder if aerospace controls could be used in their company's products. Thus the dividend here came in the form of a concept rather than actual hardware. The NASA material, explained Jerry Zahorsky, a research engineer at Hussman, "gave us the *idea* for the solid state approach, a pioneering thought for refrigeration people. That launched us into a study of just when solid state devices would be economically applicable to our industry." When the market proved opportune, Hussman began introducing solid state controls into its refrigeration products, with a resulting improvement in efficiency and a decrease in costs.

Many potential dividends from the space program are never realized for the benefit of industrial and general consumers—or if they are realized, are not advertised as such. Some firms, of course, will sit on an innovation, hoping to keep knowledge of it from competitors, present or future, while other firms believe that their images will be better enhanced if they publicize new products as stemming from their own research activities. More commonly the failure to realize dividends is simply a matter of faulty communications. The company that has made a technological advance of some sort in the course of its aerospace work either does not see the potential nonaerospace application of what it has done or, if it does appreciate the potential, does not bother to follow through and develop it—perhaps because the product application is too far afield from its main line of business; meanwhile another company that might be able to make use of the innovation never even learns about it. NASA is attempting to improve the dissemination of technical information throughout industry, but there is no doubt that many, possibly most, dividends fall by the wayside simply because not enough people know about them.

Various legal, political, and social factors also mitigate against the transfer of aerospace technology from laboratory to marketplace. The use of new materials in construction, for example, is restricted by present building codes, while in such fields as transportation and ecological monitoring and control, the implementation of major programs affecting large areas is impeded by the political fragmentation of the nation into many relatively autonomous divisions and subdivisions. And probably most important of all is the basic human tendency to follow the tried and the true, to resist change until there is no alternative except to change.

Even when all factors combine in such a way as to apparently ensure the successful transformation of an aerospace product into a consumer item, failure may still ensue. For example, in at least one case—a paint developed to protect satellites from the harsh environment of outer space—the new product failed in the marketplace because it was literally too good to be true! The paint had seemed ideal for house exteriors; after all, any paint that will withstand 800 to 1,000 scrubbings and still retain a semiporcelain gloss should hold up well through rain and snow. Nevertheless sales proved dismal when a Chicago firm, United Coatings Company, began manufacturing it for the consumer market. According to Alexander Yacek, company operations manager, "The people just couldn't believe you could buy something that good. So nobody did. We were even issuing a twenty-year guarantee—but people thought it was a gimmick." Painting contractors, too, were less than enthusiastic about a product having such a long life.

Despite all difficulties, the dividends from the space program already are great, and they are bound to increase tremendously in number and in value in the future. As an example of the way in which dividends are developed and applied in nonaerospace fields, consider what has happened in one small but discrete section of NASA's research activities: the development of new or improved fireproofing materials.

As a result of the Apollo spacecraft fire that took the lives of three astronauts in 1967, NASA began an arduous search for nonflammable materials for use in its manned spacecraft. More than 33,000 items were tested or developed, of which fewer than 1 percent eventually were accepted by the space agency.

NASA's standards were especially severe because of the particular conditions with which it had to deal. The atmosphere of the Apollo spacecraft was 100 per cent oxygen, and most nonmetallic materials will burn in pure oxygen. The spacecraft, also, was a closed system, meaning that any odors and toxic gases produced by combustion would be recirculated along with the oxygen atmosphere. On the other hand, the atmosphere within an automobile, an airplane, or a home contains a mixture of oxygen and nitrogen, and such a system is an open (ventilated) one.

Because of these differences, many of the fireproofing techniques and materials that failed to meet NASA's standards for use in manned spacecraft still are eminently suitable for use in homes, schools, hospitals, airplanes, automobiles, boats, and trains. Materials in this category include fibers and fabrics, paints and coatings, and a variety of plastics. A West German firm, Papierfabrik Scheufelen, even produced a paper that will not propagate a flame when ignited, although it does carbonize, or turn to ash; the paper has since been used by United States Playing Card Company to manufacture fireproof cards for crews in high-pressure decompression chambers, submarines, and similar environments.

Not all the products of NASA's testing program were new. Asbestos and glass fibers, for example, have been used for many years for fireproofing. However, both tend to shed excessively on account of their short staple length, and therefore neither could be used in pure form in the closed system of the spacecraft, where the detached particles would drift about in the cabin atmosphere. NASA engineers compensated for this deficiency by utilizing the fibers in composite layups with other materials, so that the shed particles would be contained in the structure of the insulation. The outer layer of the Apollo spacesuit, for instance, is composed of fiberglass cloth that has been coated with Teflon and reinforced at wear points with cloth patches made of very fine chrome steel or stainless steel wire. Other fabrics tested and found to be nonflammable in oxygen were polybenzimidazole, Durette, and kynol, the last of which can be processed into felts and battings that retain their identity even when exposed to flame temperatures as high as 4,500° F.

The widespread use of such fabrics in nonaerospace applications may be restricted for the time being by their high costs, but already they have proven valuable in certain specific fields. For example, the Space Division of North American Rockwell Corporation, which developed a fiberglass cushioning material for couches in the Apollo spacecraft, has granted a manufacturing license to the King Koil

Following the three Apollo 14 astronauts as they enter the van that will take them to the launch site is a fireman wearing a specially developed suit and helmet. The protective clothing is being tested by fire departments nationwide. (NASA)

Sleep Products Division of United States Bedding Company, which intends to incorporate the padding in a variety of products, including hospital mattresses for use with oxygen tents.

Fire departments and civil defense units, meanwhile, could make use of the several "fire entry garments" that NASA has devised for the rescue crews that stand by during training and actual launching operations. A multilayer garment of Durette and Fypro has been evaluated with promising results by the Houston Fire Department. Similarly, hospital gowns, bed linens, and mattress covers made of fireproof Beta cloth have been tested by the M. D. Anderson Hospital and Tumor Institute in Houston. The gowns and linens were designed for use in the hospital's hyperbaric oxygen chamber, where the atmosphere is 100 per cent oxygen at three times sea level pressure.

Such fireproof fabrics should have even wider application in everyday wearing apparel for the public, with the possible saving of many lives. Statistics show that some 250,000 burn cases a year—30 per cent

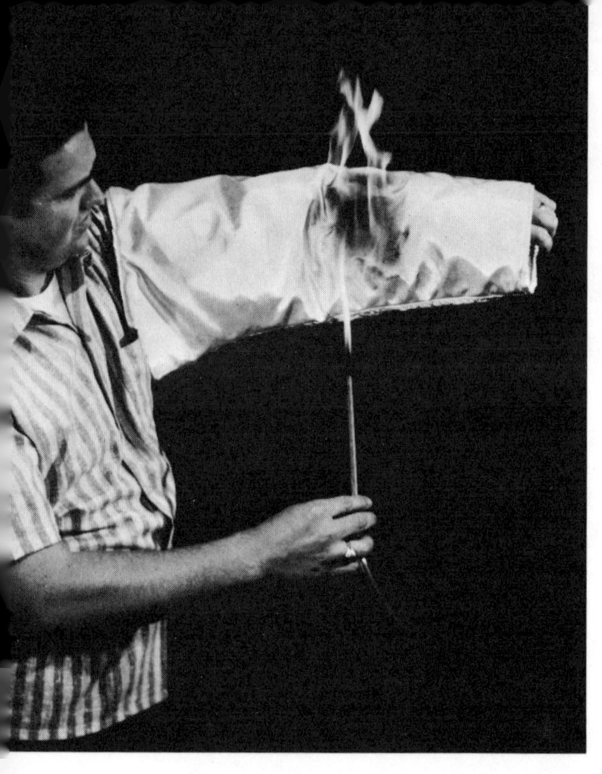

Demonstrating the fire resistance of Beta cloth, a NASA technician applies a Bunsen burner to the sleeve of an insulated, fireproof suit. Hospitals can use this cloth for gowns and linens in oxygen chambers, where the threat of fire is especially great. (NASA)

of all hospital admissions for burns—are the result of fabric fires and that from 3,000 to 5,000 people die each year on account of flaming fabrics.

In addition to fabrics, NASA tested or developed many paints and coatings, including several intumescent compounds that swell or blister when exposed to flame. The blisters contain a gas that insulates the surface beneath, thus retarding its ignition. Such coatings have been adopted by the Department of Defense for painting high-explosive bombs and the interiors of aircraft carriers. The time delay afforded by the intumescent coating gives personnel a chance either to remove the bombs from the burning area or to evacuate the area. The advantage of using such paints on storage tanks for natural gas, gasoline, or other flammable materials is obvious.

Perhaps the most versatile fireproofing material to stem from NASA's research is a specially compounded fluorocarbon elastomer commonly called by the trade name Fluorel or Viton. It can be foamed, cast, molded, or extruded; and it can be made up as a paste, liquid, or spray solution. There are 33 different items or uses of this substance in the Apollo spacecraft, including oxygen hoses, boot soles, headrests, gaskets, gloves, and circuit breaker cases. Since the compound will take pigmentation, it can be produced in a variety of colors, which should make it of esthetic value to decorators in fireproofing interiors of homes, offices, schools, and hospitals. As a coating on electronic circuit boards and components, it could materially increase the safety of television sets, radios, and record players. Composite layups of several nonflammable materials coated with the compound were made by NASA into attractive interior paneling for commercial aircraft on an experimental basis. The technique has been adopted by several airlines, and NASA itself is using the material in its executive aircraft.

In another area of fireproofing research, NASA's Ames Research Center, in looking for lightweight plastic foams for use in reentry heatshields for spacecraft, developed a material with great promise for industrial fire protection, particularly against fuel fires. The material is an extremely lightweight polyurethane foam with special additives. When exposed to flame it forms a tough, protective layer of char while at the same time releasing gases that help to quench the fire by preventing oxygen from reaching it. Used as an insulation in the walls of houses, it would make the home more fireproof. By adding glass or carbon fibers, the collapsing strength of the foam can be increased from its normal 30 pounds per square foot to as high as 210 pounds per square foot. Thus the material has a potential for use in automobiles, trucks, pleasure boats, and trains as well as such industrial facilities as oil refineries and chemical plants.

An additional dividend for industry as a whole that stems from NASA's fireproofing research is the development of testing procedures and standards for odors and toxic gases generated by burning materials. Finding that there were no standards or testing procedures in this area of research, the agency devised its own tests and created an arbitrary hedonic odor scale ranging from "irritating" through "'undetectable," with a corresponding numerical scale from 4 to 0. For spacecraft purposes, a material had to have an index of 2.5 or lower. The American Society for Testing Materials took the procedure under evaluation as a potential national standard. With so much attention being given to the quality of the natural environment, testing of this sort should be useful for a wide range of consumer products.

BETTER PRODUCTS FOR EVERYDAY LIFE

Besides building the rockets to take men to the Moon and back, NASA had to pioneer many new techniques for feeding, clothing, and sheltering astronauts on their long voyages through space. These innovations, so essential for performing the routine tasks of living and working in an alien environment, are just beginning to be translated into better products for homes, schools, and offices.

In the area of food technology, for example, NASA was concerned with developing high-density, highly nutritious foods that weigh little, occupy a minimum of space, and do not require refrigeration. It also searched for new, more efficient ways to prepare foods. These efforts have resulted in a number of dividends for the armed services, the airline industry, and the food processing and preparation industries, as well as for the housewife in the home kitchen.

Much of the research on food was done at the U.S. Army's Natick Laboratories in Massachusetts, spurred by NASA's requirement for a freeze-dehydrated food that could be prepared in 10 minutes. Consequently the Army's combat field rations were improved so that they could be readied for eating in just 5 minutes, compared with 20 previously. In addition the Army developed a collection of 32 foods that are compressed into a volume slightly larger than a shoebox. Weighing only 10 pounds, each of these nutritional "building blocks" contains such foods as chicken, beef, potatoes, tuna, bacon, cheese, rice, and gravy, for a total value of 22,000 calories—enough energy to keep one individual going for 10 days. The compressed meals have great potential for use in civil defense and disaster relief. Had they been available in quantity, they would have been ideal for relieving the famines in Biafra in 1968 and East Pakistan in 1970. Because of their high nutritional value in relation to weight, the building-block meals also should prove valuable on exploring expeditions and at research stations in remote locations in the Arctic and Antarctic.

Food processors began to explore the commercial possibilities of the freeze-dehydration process once it had moved from the laboratory to the pilot production line for space foods—with the result that grocery stores now carry such products as coffee, tea, soup, potatoes, and onions in freeze-dehydrated form. Utilizing the process in combination with packaging techniques developed for space foods, several cereal companies offer dry breakfast foods that include freeze-dehydrated fruits and berries, which reassume their original color, texture, and form when milk is added to them. In addition, Pillsbury Company is marketing Space Food Sticks, which were developed originally as contingency foods that could be eaten by an astronaut without removing his space helmet. The sticks, slightly larger than cigarettes, come in a variety of flavors and furnish 44 calories apiece. Consisting of 8 per cent protein, 70 per cent carbohydrates, and 13 per cent fats, in addition to vitamins and minerals, they are sold as a quick source of energy for hunters, fishermen, athletes, and active, hungry teen-agers.

Cooking and food preservation techniques also have been improved as a result of aerospace research. Commercial aircraft, for instance,

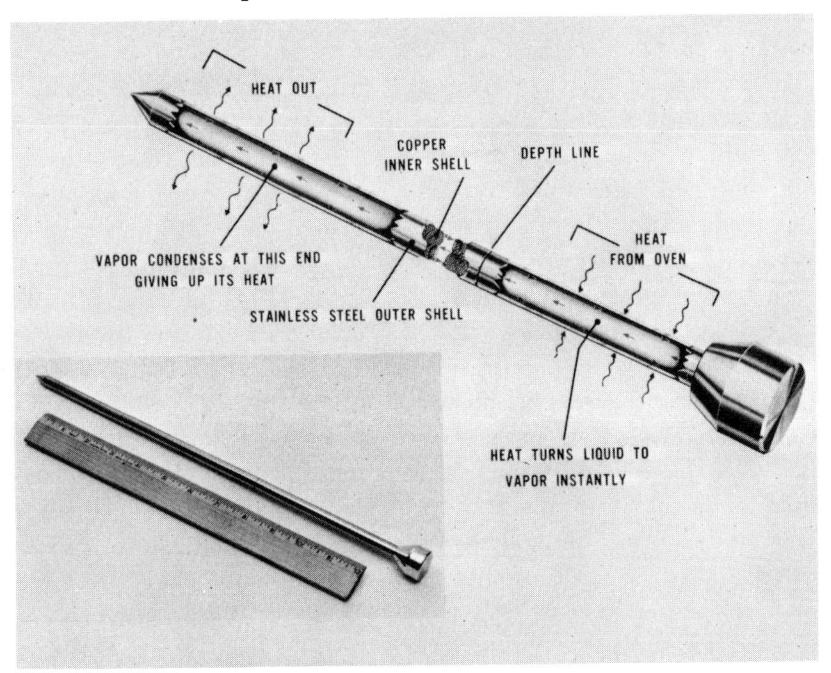

The Thermal Magic cooking pin, an offshoot of the heat pipe developed to carry heat away from electrical components on rockets, halves the time required to cook—or freeze—roasts and other meats. (NASA)

now carry lightweight microwave ovens originally developed in the space program. By using high-frequency radio waves to heat the inner part of the food—in effect, cooking it from the inside out—the microwave ovens cut cooking time from hours to minutes. Another device for internal cooking that stems from aerospace research is the "heat pipe," invented by the Los Alamos Scientific Laboratories as a means for carrying heat from electrical components and nuclear space engines. The heat pipe is a stainless steel tube containing an internal wick and a volatile fluid such as water or alcohol. Heat enters the tube at one end and vaporizes the fluid, which carries the heat toward the cold end where it is radiated away. The vapor then condenses and travels back down the wick, and the cycle starts over. In its commercial incarnation, a small version of the heat pipe called Thermal Magic is being marketed by Energy Conversion Systems of Albuquerque, New Mexico. By inserting the pipe into a fowl or roast, cooking time can be reduced by half. The pipe can be used with conventional electric and gas stoves, or, working in reverse, it can be used to "pump" heat out of meat, thus freezing it faster.

In the area of food preservation, Garrett Corporation of Los Angeles modified the cooling system it had developed for the Gemini spacesuit to make a refrigeration system for aircraft galleys that is self-contained, quiet, and requires no batteries or external power supplies. In addition it uses nitrogen rather than air, thereby providing an atmosphere in which food decay is greatly retarded. Installed in an air freight plane, it permitted a large grocery chain to fly sun-ripened pineapples from Hawaii to California overnight. Within a few weeks pineapple sales were up 40 per cent despite a price increase of 5 cents per pound.

Space research facilities have even been used to make better cake mixes for the housewife. General Mills, Inc., had a problem in testing cake mixes that was solved by a vacuum test chamber of Environ Laboratories, Inc., Minneapolis. The chamber can duplicate various altitudes, such as that of mile-high Denver. Since the cake mixes must rise at sea level as well as in Denver, the ingredients must be blended with care. By testing various mixtures in the space chamber, the company arrived at the optimum product, ensuring that the Denver housewife gets the same quality of cake as the housewife in New Orleans.

Clothing is another important area in which improvements have been made in commercial products, as a result of NASA's search for the best apparel for astronauts in Moon rockets and orbital space stations. Dividends here range from a new process for laminating fabrics, now frequently used in place of stitching in brassiere manufacture (laminating is cheaper and results in softer, more flexible garments), to such new fabrics as Astrolon, a lightweight, quick-dry cloth that is being used in T-shirts, jackets, and undergarments. Another new fabric, a lightweight woven nylon that is aluminized by the vacuum-disposition process used to produce spacesuits, traps and retains up to 80 per cent of the heat generated by the body; thus the body becomes its own heater. This fabric is being used in ski jackets and other outer garments.

Also available for the home is a more efficient blanket that stems from a search for a superinsulating material for the Apollo spacecraft. Made by Norton Company, Winchester, Massachusetts, the waterproof, washable blanket consists of a special plastic base coated with

Emergency rescue blanket, made from a special plastic material coated with 1/1,000,000 inch of aluminum, weighs only 3 ounces but is strong enough to serve as a stretcher. It is derived from material used in the Echo satellite and the Apollo manned spacecraft. (NASA)

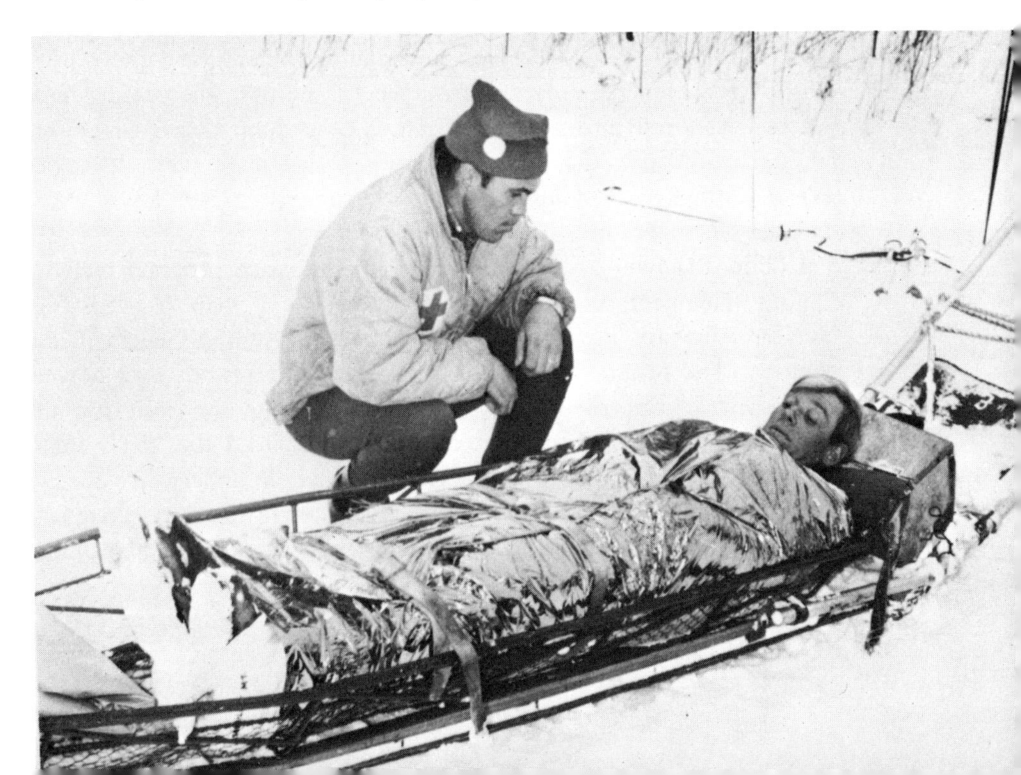

aluminum. The material is just ⅛ inch thick; a 32-square-foot blanket weighs 11 ounces and can be folded into a packet no larger than a paperback book. The general public should be attracted by the blanket's $2 retail price. In addition, the fact that the material reflects radar signals has helped to convince the Canadian Coast Guard Service, the Royal Australian Air Force, and several police departments in the United States to adopt the blanket as an official rescue item. The company is now investigating additional uses for the material in jackets and other clothing.

More comfortable houses, apartments, and offices are possible through an environmental control system built around the technology that provided the fuel cell electric power supplies for Project Apollo. United Aircraft Corporation, Windsor Locks, Connecticut, drawing on its experience in manufacturing fuel cells for the Moon landing program, plans to evaluate the suitability of a gas-powered cell for use in single-family homes as well as in apartment houses, shopping centers, and other large installations. The unit would control indoor temperature and humidity while furnishing electricity. Another power system that shows promise for both heating and cooling homes is a Radio Corporation of America adaptation of a thermoelectric air conditioning unit, developed by aerospace research for use in mobile communications vehicles. The operation of this unit is quite novel: as electricity passes through a semiconductor alloy made of bismuth and tellurium, heating occurs on one side and cooling on the other; thus, by reversing the polarity of the electric current, the cooling unit can be converted into a heating unit. The system is said to be 12 times more reliable than the conventional air conditioner. A single unit is capable of cooling a four-room apartment to 15° lower than the outside temperature.

Additional power systems derived from the space program include Conbus (Compact Nuclear Brayton System) and Snap (System for Nuclear Auxiliary Power), both developed by Aerojet-General Corporation. The former system uses a small nuclear reactor as a power source for an electric generator, thus eliminating the need for an air-breathing engine. Though it can be used on land too, the system is especially suitable for supplying electric power in underwater operations such as drilling. Meanwhile the Snap atomic batteries, designed to power instruments left on the Moon, are being used in unmanned and remote locations on Earth. One Snap unit, for instance, is providing power for the control system of an oil well in the Gulf of Mexico. Other units provide electricity for navigation lights and foghorns on offshore oil installations. Their potential use as emergency power supplies and in operations where uninterrupted long life and low maintenance are requisite seems almost unlimited.

The solar cell, whose development made communications satellites possible, also has potential terrestrial applications that are just beginning to be exploited. Though solar cells operate only during periods of sunlight, electricity generated then can be stored in batteries for use in the dark. Such cells should be particularly useful as emergency power sources for communications facilities at remote locations. For example, an emergency call system on the Los Angeles freeways and a telephone system in South Africa already make use of solar cells. They also have excellent potential as power sources for pipelines and pumping stations, especially in desert or semidesert regions, which have little cloud cover.

As far as the daily life of the individual consumer is concerned, one of the most important long-term effects of the space program may be the impact that it is having on the design of the home in which he lives. As a result of their experience in planning life support systems for astronauts, engineers are rethinking the concept of what a house should be. For instance, Grumman Aerospace Corporation, Bethpage, New York, builder of the lunar module, is developing a plan for tying such household subsystems as heating, cooling, cooking, water supply, and sewage and refuse disposal into one integrated unit. The idea is to use the unwanted products of one subsystem to create useful constituents for another. Thus waste heat from the furnace, hot water heater, washer, or drier would be used in a low-temperature water evaporation unit rather than being vented into the air. This unit, in turn, would reprocess waste water from the washer, bathtub, kitchen sink, toilet, etc., for reuse. Water from the city supply would be used for drinking and cooking only. An incinerator in the system, again using waste heat, would reduce garbage, refuse, and hygienic wastes and produce energy for the heating and air conditioning units. Integrated systems of this sort, besides lowering the householder's utility bills by making more efficient use of the water and electricity he purchases, would, if adopted on a large scale, help to reduce the nation's refuse disposal and air pollution problems while minimizing demands on portable-water supplies.

Ultimately, the house itself may be built of a material developed originally for rocket motor cases, as a result of a joint project of the University of Michigan's Architectural Research Laboratory and Aerojet-General Corporation. They produced a prototype of a house for about 11 dollars per square foot—considerably lower than the cost of conventional housing—using fiberglass-reinforced plastic and filament winding techniques developed by the firm for making motor cases for solid-propellant rockets. A home of this construction is on display at Disneyland as "The House of the Future."

The same material has been used by United Aircraft Corporation, Riverside, California, as piping for sewage, irrigation, storm drains, and water mains. Unlike concrete mains, which may lose through evaporation and leakage as much as 80 per cent of the water they transport, the fiberglass-filament pipes lose none. The pipes are particularly useful in industrial applications because they are impervious to corrosive liquids. Moreover, they bounce instead of shattering or chipping when dropped from the back of a truck and, since they weigh less than concrete or iron, are cheaper to ship and can be handled with lighter equipment.

The versatility of the fiberglass-filament process also is illustrated by its successful use in the fabrication of aircraft wings and a variety of high-pressure storage bottles for aircraft, ships, and submarines. In addition, the technique of filament-winding large cylinders has been applied by Lamtex Industries, Inc., in manufacturing railway tank cars of 8,000-gallon capacity which weigh 5 tons less than a conventional steel car.

AEROSPACE INSTRUMENTS COME DOWN TO EARTH

The civil engineering profession and the construction industry have benefited generally from several aspects of space research and technology. In San Diego, for instance, architects and builders borrowed ideas from the launching pads at Cape Kennedy and the huge spacecraft tracking antennas at Goldstone,

California, to design a complex of nine circular apartment houses, 18 to 24 stories tall, that will revolve once every 3 hours on a central service axle, giving occupants a panoramic view of the city. And in England architects drew on studies in aerodynamics, made using model homes and buildings in a wind tunnel, to design shopping centers in Leeds and Croydon in such a way that unpleasant wind eddies would be minimized. Similar studies in Canada have led to a proposal by Prof. B. Etkin of the University of Toronto for using "curtains" of air like walls to protect areas from rain and wind.

A better knowledge of how dams react to stresses of earthquakes, as well as to internal dynamic tensions, is now available from data gained in California, where stress-monitoring cells developed by Aerojet-General for measuring the thrust of rocket engines have been embedded in the Oroville and Castaic dams. The same company also has adapted an ultrasonic testing device, developed to detect flaws in rocket motors, for use in locating structural flaws in masonry buildings such as schools and hospitals. When tested by the California Office of Architecture and Construction, the instrument detected not only flaws that had been intentionally built into six simulated wall panels, but also those that had occurred spontaneously in the fabrication process. The technique is much more reliable than the old method of taking core samples from completed buildings.

In like manner, a gauge built by the Marshall Space Flight Center, Huntsville, Alabama, to measure the thickness of paint on metal surfaces and of the insulation foam coating on rocket propellant tanks has been adapted for measuring the depth of freshly poured concrete in roadbeds. It can measure thicknesses up to 1 foot to an accuracy of less than ¼ inch. The same instrument also may be used by airlines to scan passengers for hidden weapons, since it can be adjusted to detect large masses of metal while ignoring rings, pens, and small jewelry.

The adaptation for nonaerospace uses of instrumentation developed by space technology is only beginning to be realized. Several dividends of this type are, however, already close to actuality.

1. Radiometers installed in weather satellites and space probes have been modified for use on Earth to measure the depth of snowpacks. Experiments on Mt. Rainier indicate that the truck-mounted instruments provide much more accurate measurements than obtained by

Not much larger than the coffee cup on which it rests, this sensor measures wind velocities as faint as 1/10 foot per second. Developed primarily for use with short take-off and landing aircraft, it also has been used to make critical airflow measurements in coal mines. (NASA)

rangers on foot using hand-operated coring devices to take samples. With better data, it should be possible to make more reliable predictions of the amount of water that will result from spring thaws and, hence, the amount of hydroelectric power that can be generated.

2. A sensor designed and developed by NASA for measuring the speed of V/Stol (Vertical/Short Take-Off and Landing) aircraft shows promise in the field of mine safety. Engineers have used the sensor to measure low-speed air movements in coal mines, to make sure that enough fresh air is being supplied to the coal face to dilute and carry off explosive gases and dust particles. NASA also has suggested that a laser beam operated in conjunction with the sensor could detect such poisonous gases as methane and carbon monoxide.

3. Barnes Engineering Company, Stamford, Connecticut, has modified the infrared Earth horizon seeker that it built for the Mercury spacecraft so that the device can measure the thickness of a red-hot steel rod continuously being formed in a steel mill. Even though the rod travels at speeds of up to 75 mph while being cast, the sensor measures its diameter to an accuracy of a few thousandths of an inch,

greatly simplifying the problems of quality control. The device seems applicable to such related processes as the manufacture of glass rods and tubes.

4. Since 1968 the U.S. Army Corps of Engineers has utilized space-developed sensors and telemetry to monitor water levels of crucial rivers and streams in Alaska, California, and Massachusetts. Commenting on the system's operation in Alaska—where it has given as much as two days' warning of an impending flood—the Corps reports:

> The highly reliable, inexpensive reporting system can prevent millions of dollars of flood damage. The central station installation ran about $20,000 and the remote stations were $6,000 each. Six radios cost $1,000 each. Using data from this system, one man and the printout station can assess and forecast the danger faster and more accurately than a team of ten men at various locations in the watershed, assuming weather permitted such a team to get into the remote region.

NEW MATERIALS AND TOOLS FOR INDUSTRY

In order to construct spacecraft and rockets, the aerospace industry had to develop a host of new materials and new tools. The rigors of the environment beyond Earth's atmosphere ruled out many metallic and nonmetallic materials used in aircraft, ships, and automobiles—they simply would not withstand the harsh vacuum conditions, temperature extremes, and blistering radiation from the Sun and deep space. New metal alloys, adhesives, plastics, and insulating materials were required. At the same time, new materials and new tasks called into existence a wide variety of new tools, ranging from extra-large welding rigs 33 feet in diameter to small, hand-operated pliers for stripping wires more efficiently. Today, as a result of these many developments, American industry as a whole has a far greater technological potential than it did when the space program began.

The impact of space technology on industry generally is well illustrated by what has happened in the field of metallurgy. Whereas only a few high-strength aluminum alloys existed at the beginning of the space program, there is now a large family of extremely strong, weldable aluminum alloys that have proved useful in a number of commercial applications—milk tank trucks, for instance. Steel manufacture, too, has benefited from space research, as has the usefulness of many rarer metals. Today there are available maraging steels that require a 900° F treatment to develop full strength, as well as extremely hard stainless steels that will hold up to temperatures from 150° F to 600° F. Meanwhile the search for a lightweight tank for liquid hydrogen led space engineers to develop a titanium pressure vessel that is now used in the chemical and petroleum industries and has useful applications in military and civilian aircraft. Similarly, new markets should open up for tungsten as a result of the discovery in the course of aerospace research that it can be made ductible at room temperature through the addition of a small amount of rhenium. NASA also pioneered in the technology of beryllium, developing new methods of alloying and machining this lightweight yet extremely strong and heat-resistant metal.

New plastics and adhesives also have appeared as a result of space research. Hystil, a plastic that is stronger than steel but not as stiff, proved unsatisfactory for its intended application in rocket engine nozzles, but the firm that developed it—TRW, Inc., Redondo Beach, California—has joined with Commonwealth Oil Company, New York, to exploit what seems to be a wide range of commercial possibilities for the product. Automobile manufacturers are interested in using it for bodies and frames; other potential uses include piping for corrosive chemicals and casings for electronic components.

A family of polymers called pyrrones also has been developed. Much more resistant to radiation than previously available polymers, they have outstanding stability at high temperatures and can be tailored with a wide range of electrical properties. They should be useful in the electronics and aircraft industries. In addition, NASA studies of the outgassing of plastics and rubbers have revealed many of the impurities in such materials. Identification of these substances has proved valuable to industries that have stringent specifications for polymers, such as food processing and automotive component manufacturing. Meanwhile polymer resin adhesives, incorporated on the Apollo spacecraft and Saturn rocket, are finding a nonspace market. The automotive industry is especially interested in them for such purposes as securing body trim. The racing car that won the 24-hour

The same polyurethane foam material used to insulate the second stage of the Saturn 5 lunar rocket is applied between the fish well walls of a tuna clipper, at right. With the foam insulation, temperature of the fish compartments can be kept near 0°F. (NORTH AMERICAN ROCKWELL CORP.)

Le Mans race in France in 1967 was entirely bonded by adhesives. Several Detroit manufacturers are planning to introduce explosion-proof reinforced-plastic gas tanks in cars during the 1970s.

A plastic foam insulation for the propellant tanks of the Saturn 5 rocket now finds use in commercial fishing boats. Developed by North American Rockwell Corporation, the polyurethane substance is a more efficient and longer-lasting insulator than cork. Commercially, the foam, which is applied to metal structures by spraying, has been used to insulate the holds of a tuna ship built by Campbell Machine Company of San Diego. The insulation will keep the temperature of the fish compartments near 0° F. Another application foreseen for the foam is in refrigerator cars for trains and trucks. It also has a potential use in the automobile of the future, which may operate on propane or liquefied gas; a fuel tank so insulated could help to maintain the–160° F temperature required.

Space technology is responsible for a new breed of extremely high-strength structural materials called composites. In addition to fiberglass-reinforced plastic, the composites include fibers of carbon or boron mixed in an epoxy resin. Carbon-epoxy composites are used in the blades of compressor fans in jet engines, and boron-epoxy composites in helicopter rotor blades as well as landing-gear doors and rudders of aircraft. A boron-epoxy laminate also is used to add fatigue strength to the wing pivot mounting of the F-111F jet fighter. Prepared in sheets and bonded to the steel structure with a special adhesive, the composite strip doubles the metal-fatigue life of the pivot, cuts down on maintenance costs, and greatly increases safety. Other materials undergoing tests for use in composites include filaments and fibers of aluminum oxide, silicon carbide, boron carbide, beryllium, and tungsten.

Another innovation in materials technology is the honeycomb-core composite, a product of the aerospace industry's search for lightweight but high-strength construction materials. The composite panels —thin sheets of metal bonded to a honeycomb core of aluminum or plastic—can be used to make house trailers that are 45 to 50 per cent lighter than those of conventional construction. This type of panel also is finding its way into prefabricated homes. In addition to their light weight and high strength, such panels are practically maintenance-free and are easily replaced.

New tools developed through the space program also are making the transition to the general industrial market. Typical of the small tools that have been adopted for nonaerospace applications is a NASA-devised coaxial-cable stripper and cutter, now being manufactured commercially by Western Company, San Clemente, California. Its use has resulted in savings of up to $5,000 annually for some firms. Similarly, Aerojet-General has begun commercial production of a 3-pound heat source that it developed as a means of joining liquid-propellant lines in rocket engines. The device provides instantaneous fumeless and flameless local heat to weld tubes and pipes of up to 1 inch in diameter. It has many potential applications in the air conditioning and refrigeration industries. Another hand tool developed by NASA uses the explosive power of a 22-caliber blank cartridge to join two sections of tubing in a swaging operation. It is being produced for the commercial market by Newport News Shipbuilding and Drydock Company in Virginia.

Perhaps the oddest of the new tools derived from the aerospace effort is one that is made of ice. It was developed by Boeing Company engineers at the Kennedy Space Center in Florida as a means for spreading and molding adhesive. A small handle containing a coil through which refrigerant circulates is inserted in a water-filled mold, and the ice tool is formed in much the same way as an ice cream stick. Since the adhesive does not stick to the ice, it is possible with this tool to work the material into areas where performed seals cannot be used. The tool also is useful in applying sealants to nonporous surfaces such as glass.

From research into the effects of micrometeoroid impacts on spacecraft has come a new tool for the mining and construction industries. To calibrate the monitoring instruments it was developing, Exotech, Inc., Rockville, Maryland, devised a method of simulating micrometeoroid impacts using small, high-speed bursts of gas. The same technique, applied in producing a water cannon, resulted in a machine that will fire 10 bursts per second with impact pressures as great as 8,000 pounds per square inch. According to David Fain, vice president for marketing, 1 cubic foot of water will move 2 cubic feet of rock. The company feels that this machine could revolutionize mining and tunnel boring.

Other dividends illustrating the breadth of the range of applications of space-derived tools include:

1. An electromagnetic hammer, invented at the Marshall Space Flight Center to remove dents from the large curved surfaces of the Saturn 5 rocket's propellant tanks without weakening the metal. The device has been used to remove dents from helicopter blades and is being utilized in shipyards and automotive manufacturing as well.

2. A machine for testing the quality of flared tubing, also developed at the Marshall Space Flight Center. The machine has been marketed commercially by Metrophysics, Inc., Santa Barbara, California. Besides its usefulness to any industry employing flared tubing, it may prove an asset to police departments since it appears to have the potential ability to read impressions left on pages of paper pads from which the written page has been removed.

J. R. Rasquin (left) and Morris Whittington with the electromagnetic hammer they adapted for removing dents from the Saturn 5 rocket booster. An electrical shock wave generated by the magnetic hammer at top is amplified as it passes through the curved horn and is concentrated in a small pocket on the anvil below, producing sufficient pressure and temperature to synthesize diamonds from graphite. (NASA)

3. A stepping drive, developed at the Marshall Space Flight Center to position scale model rockets in wind tunnels. This invention has been adopted by a Los Angeles clockmaker, Ropat, Inc., as a means for operating a clock having a digital display rather than a conventional face and hands. The company expects the new clock to have a sales volume of $500,000 a year—and this value could be much larger if the clock is incorporated in automobiles, for which it seems particularly well suited.

SPACE TECHNOLOGY AND THE COMPUTER REVOLUTION

By no stretch of the imagination can the space program claim paternity of the computer. However, many improvements have been made in computers as a result of advances in microelectronics and solid state electronics that derive from space technology. The upgraded computers, of course, are of use to industry and business generally. Moreover, many of the computer programs developed by NASA can be adapted for nonspace projects.

For example, a NASA program for balancing the variables, or parameters, of a design in the optimum combination has been used by engineers at the Bonneville Dam to design circuits for power control, as well as by General Foods, Inc., to maintain consistency in food products. The University of North Carolina used the same program to determine how funds could be distributed among public health and medical research programs to achieve optimum results in such diverse areas as improved living standards and education. This highly flexible program also has been used in the electronics, petroleum, and chemical industries.

Flotran, a unique computer program developed for the Marshall Space Flight Center, automatically produces flow charts from many different input statements. It also has found wide usage outside the space industry. Other programs of a similar application include Gremex and Nastran, both of which are available from NASA at only the cost of reproduction.

Sometimes computer hardware developed for the space effort has utility in conventional enterprises. A Packard Bell computer for automatically checking out the Saturn 5 rocket has been adapted for railroad scheduling and oil refinery processes. Remington Rand computers developed for use in microminiaturization of electronic components for space vehicles can be adapted for translating foreign languages or for library research.

NASA's computers also store information about the huge number of technical innovations that have been made in the course of the manned and unmanned spaceflight programs. This information is available to industry at only a fraction of the developmental cost. As an example of what this can mean for a company, consider the case of William Ferwalt, who employs seven Nez Percé Indians on their Idaho reservation to make oscilloscopes.

Ferwalt wanted to know what space research had found out about oscilloscopes, and he paid $190 for a computer search of NASA's technical data bank. Among the information he received was an item for building a special type of instrument that he had not considered.

The NASA computer, at a cost of only $190, had given him information that he expects will be worth at least $100,000 in sales over a period of a few years.

Larger companies, such as Litton Industries and Alcan Aluminum Corporation, pay as much as $1,000 to $5,000 a year for the use of data banks containing material in their areas of interest. However, for as little as $80, a company with less specific needs can obtain a computer "readout." Joseph Ciarimboli, manager for technical planning of Dart Industries, Inc., an organization that makes a wide range of housewares, plastics, and drugs, periodically purchases NASA's computer service. "When we are probing new business areas for Dart, we need a quick reading to find out if a particular technology has consumer applications," he says.

FUTURE DIVIDENDS FROM SPACE TECHNOLOGY

There seems to be no end to the application of knowledge accruing from space technology. In many cases only imagination and a willingness to innovate are required to gain more practical dividends. For example, an experimental shock absorber for the couch in the Apollo spacecraft has been tested by the Bureau of Public Roads for guard rails on highways. Results show that the shock absorber reduces a 60-mph collision to the equivalent of a 5-mph impact. Ford Motor Company has undertaken a program to test the device in a special automobile bumper that would permit collisions at 5 mph with no damage to the car. The economic consequences to the owner are obvious when considering the announcement by Allstate Insurance Companies that premiums would be reduced by 20 per cent on all cars so equipped.

Engineers of the Industrial Fabrics Division of West Point Pepperell, Inc., New York, were intrigued by the commercial possibilities of the bags developed by the company for righting the Apollo spacecraft automatically if it came to rest upside down in the ocean. The idea led them into the field of automobile safety; they developed similar bags that inflate automatically upon impact to impose themselves between the passengers and the dashboard and windshield of the car. Similarly, a bulletproof tire developed by Goodyear Tire and Rubber Company, Akron, Ohio, for the U.S. Air Force Aeronautical Systems Division seems to have commercial applications in the automotive industry. The tires are filled with a special foam rubber instead of air. While this adds to the weight of the tire, it has the beneficial side effect of lowering the vehicle's center of gravity, thus increasing its stability.

A better color television tube is also possible through space research because of experiments and studies made at NASA's now defunct Electronics Research Center in Cambridge, Massachusetts (it was one of the first victims of the decline in appropriations for the space program). These investigations resulted in a phosphor that can produce colors ranging from deep russet to pale yellow. Possible ultimately is a single phosphor that could produce the total range of colors, doing away with the need for the three different phosphors currently used.

Similarly, chemical experiments by Aerojet-General Corporation indicate that a form of DDT is possible that has a "self-destruct mechanism." The company found the means of introducing a catalyst into the insecticide that can be slowly released to break up the DDT over a finite period of time, ranging from 6 hours to several months. If successful, this would be a tremendous boon to farmers the world over, especially to those in developing countries, who feel that they must continue using DDT because of its low cost and effectiveness—despite the newly discovered long-run threats that the insecticide presents to wildlife, and perhaps to man himself.

These are only a few of the possibilities that lie before us. It is reasonable to expect that during the 1970s new advances will be made in sensors, instrumentation, microelectronics, materials, data transmission, life support systems, and other technological areas. These advances in turn will provide a strong technological base for the development of additional dividends. As the following chapters show, a host of dividends already have been obtained in fields ranging from health and medicine to communications and navigation.

The view from orbit is alone of inestimable value. Satellites give man a remarkable overview of his planet—hour by hour, day by day, year in and year out. As sensors are improved, satellite data will become of increasing importance to meteorologists, helping them not only to forecast the weather but to develop a profound understanding of atmospheric processes; to geologists, in their search for new mineral reserves; to geographers, for studying the cultural features of the world and for upgrading and improving maps; to hydrologists, in making assessments of water resources, monitoring pollution, and detecting incipient floods and droughts; to agriculturalists and foresters, as they prepare inventories of food and fiber resources, study land use and soil condition, and spot blights and fires; as well as to oceanographers, as an aid in studying the enormous reaches of the seas and their fishery resources.

Prognostication is a notoriously risky business, but it seems obvious that the continued application of our newfound space capabilities will play a significant role in enhancing the quality of life here on Earth. Dividends from space research will not themselves put an end to environmental pollution, solve all the nation's urban and health problems, or revive its educational system. However, as the following chapters demonstrate, they do provide some of the necessary tools for accomplishing these tasks. And when the value of these dividends is added up, the nation's $40-billion investment in the space program begins to look like a very good one indeed. It is, in fact, a fantastic bargain.

2
DIVIDENDS FOR HEALTH AND MEDICINE

A VARIETY OF DIVIDENDS FROM SPACE research and technology accrued to the medical profession and its institutions during the decade that ended with the landing of the first men on the Moon. These include not only a number of hardware items but basic knowledge in biology and physiology. Undoubtedly the emphasis on manned spaceflight during this period accounts for many of these dividends. The fact that medical men played an important role during the formative years of the space program in both the United States and the Soviet Union may also explain why medicine benefited more than other professions or industries.

BASIC RESEARCH, SERENDIPITY, AND CATALYSIS

It is difficult to evaluate the contributions to basic knowledge in biology and physiology accruing from the space program. Since such dividends seldom receive publicity, these contributions are less well known than those that involve "hardware." Nevertheless, medical science has been enriched by aerospace medical research in areas such as the metabolism of oxygen by the human being and its clinical use at hyperbaric (greater-than-sea-level) pressures, the mechanics of the vestibular apparatus (inner ear), visual aberrations produced by acceleration, and the effects of ionizing radiation at the cellular level.

Typical of the contributions to basic knowledge in biology is that of Clarence D. Cone, Jr., a research scientist at NASA's Langley Research Center. In studying the effects of space radiation on human cells in 1969, he discovered intercellular linkages that could help explain the behavior of certain types of cancer cells. His discovery was made through the use of time-lapse motion pictures and examination of cells grown in tiny ponds where their movement was restricted. His films showed a phenomenon in which one dividing cell appears to be able to induce connected cells also to divide by transmitting a kind of chain reaction stimulation through a thin link of cytoplasm, or cell-wall material.

In the following year, Cone revealed another important discovery he had made as a result of his space research. Again working with cells and their response to radiation, he found that cells divide at a slow, regular rate when the electric charge upon their surface is high. However, when the charge is low, they rapidly break away from each other. Again his findings had great relevancy in cancer research.

Serendipity operates in space research as well as in any other investigative endeavor. Dr. J. Ken McDonald, a biochemist at NASA's Ames Research Center, was engaged in the study of the role of tissue enzymes in the response of the human body to the weightlessness of spaceflight. He discovered an enzyme (dipeptidyl aminopeptidase) with the unique ability to take long, complex protein molecules apart bit by bit. With the new biochemical tool, biological researchers can now take apart and study in detail the very long protein molecules of which much of the body is composed.

In some cases, space research acts as a catalyst rather than being clearly an inventor or innovator. A case in point is in the surgical use of cryogenic (extremely low-temperature) liquefied gases. With the coming of the space age, a great deal of research went into improving or perfecting the procedures for manufacturing, handling,

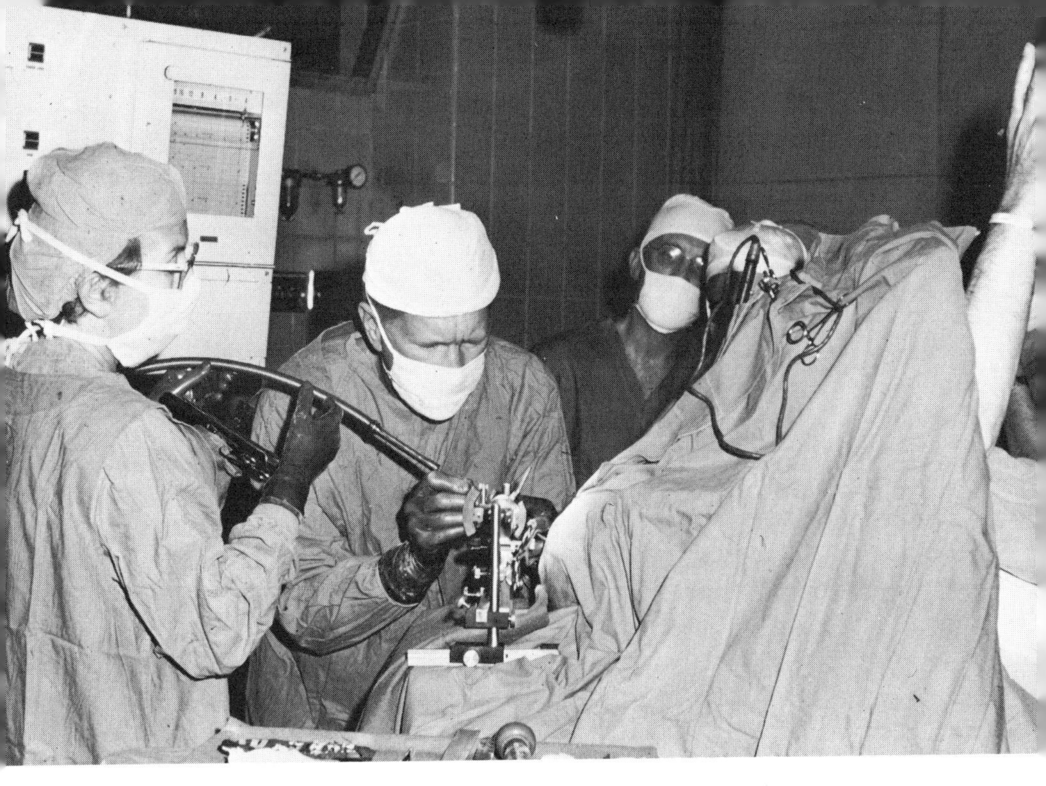

Cryosurgery—the technique of killing diseased tissues with extremely low-temperature liquid nitrogen—has benefited from aerospace research in liquid propellants. Here, Dr. Irving Cooper, one of the developers of the technique, uses cryosurgery to operate on a patient with Parkinson's disease. (DR. IRVING S. COOPER, ST. BARNABAS HOSPITAL, BRONX, N.Y.)

transporting, and storing liquefied gases such as oxygen and hydrogen.

The technique of cryosurgery was developed in the late 1950s by Drs. Irving Cooper and Arnold Lee, of St. Barnabas Hospital in New York. Instead of a conventional scalpel for cutting tissue, these surgeons invented the cryoprobe, a long needle through which liquid nitrogen circulates, producing temperatures as low as –319°F. Diseased tissue coming in contact with the cryoprobe is killed quickly and painlessly and is either removed by knife or left in place to be absorbed by the body. Cryosurgery is especially useful in brain surgery.

To improve the cryoprobe, Dr. Cooper needed to know with a high degree of accuracy the temperature at the point of the instrument. Here space engineers came to his assistance. They had earlier developed an extremely small thermocouple (electronic thermometer) that could easily fit inside the cryoprobe. So modified, the instrument now permits the surgeon to push into the brain at intermediate temperatures until the needle is in the desired location, and then lower the temperature to the minimum.

Cryosurgery has a wide range of uses and is especially employed in relieving the tremors associated with Parkinson's disease. It is also used in surgery of the prostate gland, cervix, uterus, and rectum as well as in removing tonsils.

BIOINSTRUMENTATION

One of the earliest needs of the manned spaceflight program was to take certain physiological measurements at extremely long distances. How could a doctor at Cape Kennedy or in Houston take the temperature or pulse of an astronaut orbiting 120 miles above the Earth, or walking on the surface of the Moon some 245,000 miles away? The answer is, by telemetry—the process of taking measurements at distances and sending the information by wire or radio.

For use with astronauts in space, some form of radio telemetry was indicated, and a special form called biotelemetry was developed. Prototypes were used on animals such as the chimpanzees Enos and Ham and the monkeys Able and Baker, who preceded man into space in the late 1950s.

Some of the instrumentation techniques developed during the manned space programs, such as Mercury, Gemini, and Apollo, found their way into the medical school, research institution, doctor's office, and hospital—even into the ambulance hurrying the critically ill or injured to the hospital. The potential for space bioinstrumentation is almost unlimited.

The ability to take constant measurements of the temperature, pulse, breathing rate, heart action, and brain waves of a patient in a remote location has many applications in the modern hospital. A patient isolated with a very contagious disease is a prime example. Another is the patient in a hyperbaric chamber or iron lung. Even the patient on the operating table is isolated in a sense; it is not possible to remove sterile coverings to make measurements throughout the operation. A patient being moved from one hospital to another or even from one part of a hospital to another is yet another good example of how space-age biotelemetry can be used to good advantage.

Biotelemetry also permits more frequent collections of data than would be economically possible by conventional means. A small sensor attached to the body can continuously transmit information for storage in a computer or display on a console. A doctor or nurse sitting at a central console can monitor many patients by flipping a switch. Such a system is much more economical and accurate than visiting the patients and manually taking temperatures and pulses. A start in this direction has already been made at the Presbyterian Hospital of the Pacific Medical Center, in San Francisco, where patients in an intensive-care unit are fitted with bioinstruments that transmit data on heart rate, breathing rate, blood pressure, and temperature directly into a computer. In addition to observing these data on an oscilloscope the doctor can have a 24-hour record of them printed out at the touch of a button.

Electronic biotelemetry proves very accurate compared with the human being, who may not hear a heartbeat or two through a stethoscope, may misread a column of mercury on a sphygmomanometer or thermometer, or may record readings in a script hard to decipher and on a form that must be read, interpreted, and rerecorded by another human, also subject to a variety of errors.

The uses of bioinstrumentation developed in aerospace medical research are many and various. Some can be applied directly, while others require some degree of modification.

A special sensor for the Apollo spacecraft that measured the force of its impact on landing in the ocean now finds use in fitting artificial limbs. Smaller than a dime, it is placed inside the limb and permits pressure readings to be made at several points on the patient's stump. With this information, better-fitting limbs result. Used in a special shoe, the same sensor measures the pressure on malformed feet and provides information for the design of better orthopedic shoes.

British aerospace research has produced a similar example of medical fallout. The Instrumentation Section of the Royal Aircraft Establishment, at Farnborough, came to the assistance of medical researchers at the Royal National Orthopaedic Hospital by developing a special pressure sensor that permits orthopedic surgeons to

Miniature pressure sensors: Data from the sensors in the sole of the sandal are broadcast by the transmitter in the heel, enabling orthopedic surgeons to measure loads on arthritic joints. At right, sensors that are small enough to measure fluid pressure inside the eyeball utilize techniques developed for aerospace microminiaturized circuits. (ROYAL AIRCRAFT ESTABLISHMENT; DR. CARTER C. COLLINS, SMITH-KETTLEWELL INSTITUTE OF VISUAL SCIENCES, UNIVERSITY OF THE PACIFIC)

measure the loads on arthritic joints, enabling them to follow the progress of the disease and the effectiveness of treatment. It also supplies data useful in designing artificial hip joints. The pressure sensor fits into the sole of a sandal; an amplifier and transmitter are made into the heel. As the patient walks about in the room, continuous signals are picked up by an antenna around the edge of the room.

The Bruce Lyon Memorial Research Laboratory, of the Children's Hospital Medical Center of Northern California, at Oakland, found a new use for a pressure sensor developed at NASA's Ames Research Center. Using a transmitter designed originally to send electrocardiograms of human subjects being tested on a centrifuge, the researchers and the NASA engineers adapted it for use with patients who have had a tracheotomy–an opening cut in the windpipe to facilitate breathing.

Such operations require the insertion of a small pipe, through which the air flows into the windpipe. Should this pipe become clogged by mucous or other matter, the brain might be deprived of oxygen for a period long enough to damage it irreparably or to suffocate the patient. Prior to the development of the sensor, such patients required almost constant visual care to insure that the tracheostomy tube remained clear. Now, however, a patient equipped with the new device can free the nurse for other duties. A small thermistor, or temperature sensor, from the device constantly measures the changing temperature of the cool air coming into the pipe and the warm air flowing out of it. This temperature change is converted into a signal that is transmitted by the unit's very small radio to the nurse's desk as a continuous warbling tone or as a trace on an oscilloscope. A solid state timing device on the nurse's desk monitors the incoming signal continuously. If it fails to receive the signal for 10 seconds or if it notes a decline in the volume of the signal (indicating reduced breathing), the device turns on a loud buzzer, sending the nurse to the patient's room to check the pipe.

Because of its small size (about half that of a small matchbox) and weight (less than an ounce), the unit is especially useful with children. It also allows adult patients to move about in their rooms, since the signals can be picked up anywhere in the room.

A sensor designed by NASA to detect the impact of micrometeoroids on spacecraft also finds use in medical research. This device consists of thin rods of piezoelectrical crystal (one that generates electricity under pressure) anchored firmly to the housing of the sensor on one end and provided with a flat target plate on the other. When particles hit the target plate, they bend the crystal slightly, producing a current flow in it that is proportional to the size and force of the impacting particle. This current is then con-

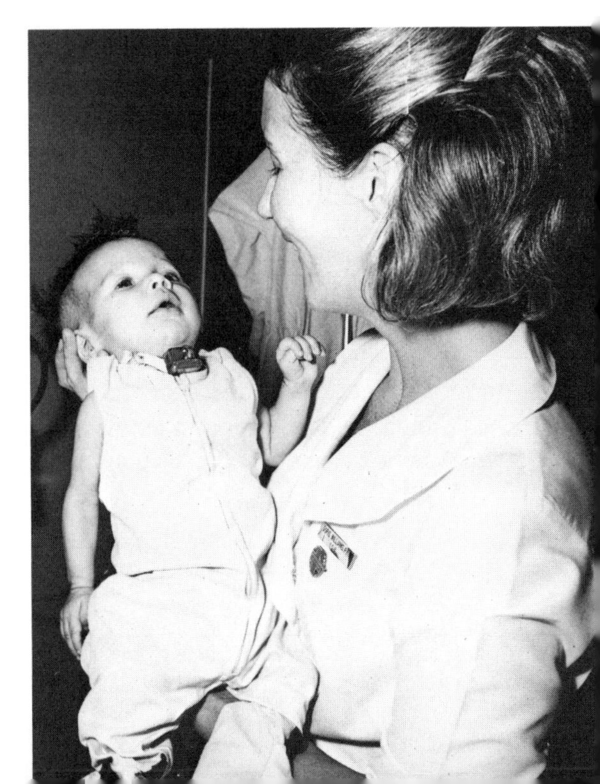

Four-month-old girl recovering from a tracheotomy wears a tiny sensor that sounds an alarm if she has trouble breathing. The device, originally designed by NASA to radio electrocardiograms from people being tested in centrifuges, eliminates the need for continually watching patients, thus freeing hospital attendants for other duties. (NASA)

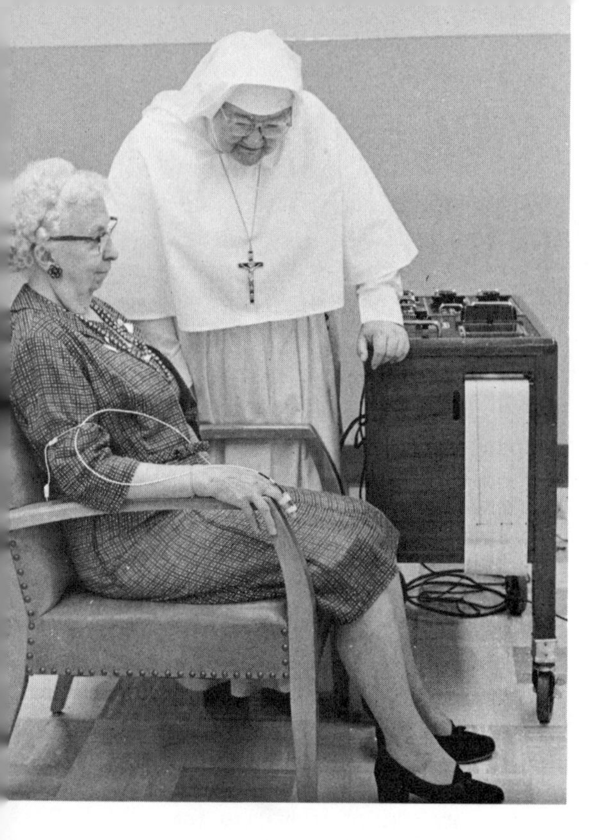

A muscle accelerometer, adapted from a sensor for detecting micrometeoroid impacts on spacecraft, is demonstrated at St. Louis University's School of Nursing and Health Services. By measuring minute muscular tremors, the device helps doctors to make early diagnoses of such neurological ailments as Parkinson's disease. (NASA)

verted to a radio signal and transmitted. The unit is so sensitive that it can detect a grain of table salt dropped from a height of only 0.125 inch. It is being used by the U.S. Food and Drug Administration to measure the effects of various drugs on the heartbeats of chicken embryos. Its advantage in this application is that the eggshell need not be pierced to install the sensor.

Another sensor, used in the Ames Center's wind tunnels to measure pressures at various points on small-scale models of advanced aircraft designs, also proves a boon to medical science. The sensor is extremely small, only 0.05 inch in diameter—smaller than the head of a pin. In fact, it is so small that it can be inserted into a standard No. 17 hypodermic needle. Injected by the needle into an artery in the groin, it is passed through the arteries into the left ventricle of the heart to measure pressures. Because of its very small size, this sensor is especially suitable for diagnostic work with babies. The sensor itself causes no pain, and it is made of materials that are compatible with blood.

An even smaller blood pressure sensor deriving from space research is one developed from a semiconductor transducer originally devised at NASA's defunct Electronics Research Center. Only 0.02 inch thick, it draws less than 0.000005 watt of electricity. The potential of the device for medical uses, especially when implanted in the body, has been demonstrated by a medical team headed by Dr. Bernard Town, associate professor of cardiology at the Harvard School of Public Health.

One of the most valuable spinoffs from space research in the field of bioinstrumentation grew from the need to provide high-speed test pilots and astronauts with body sensors that would not seriously interfere with their flight duties. The "spray-on" flexible sensor was the result. The specifications presented to NASA's researchers and Spacelabs, Inc., seemed almost impossible to realize:

1. Neither the electrodes nor their wires should be felt by the wearer.
2. Electrodes must be resistant to motion, once secured, to prevent spurious signals.
3. There must be no skin irritation resulting from frequent use of the electrodes.
4. The electrodes must not require shaving of body hair prior to emplacement.
5. Electrodes must be applied in 30 seconds or less.

The problem was solved by using a mixture of household cement, powdered silver, and acetone applied by a spray gun. The electrode so formed is less than 0.002 inch thick and the connecting wire is approximately the diameter of a human hair. Electrodes can be sprayed on in about 20 seconds, and the wires attached in about 3 minutes.

The Schaefer Ambulance Service, of Los Angeles, on hearing of the spray-on sensors, expressed interest in a biotelemetry system that could be applied by a technician while the patient was on the way to the hospital. In less than 2 minutes, the company's medical technician can now install the sensors and attach them to an electrocardiograph (ECG) which transmits heart signals over the ambulance's shortwave radio to the company's communications console. From there, the signal is sent by telephone to the emergency room of the University of California Center for Health Sciences, where it is displayed on a conventional ECG recorder for the doctor awaiting the ambulance.

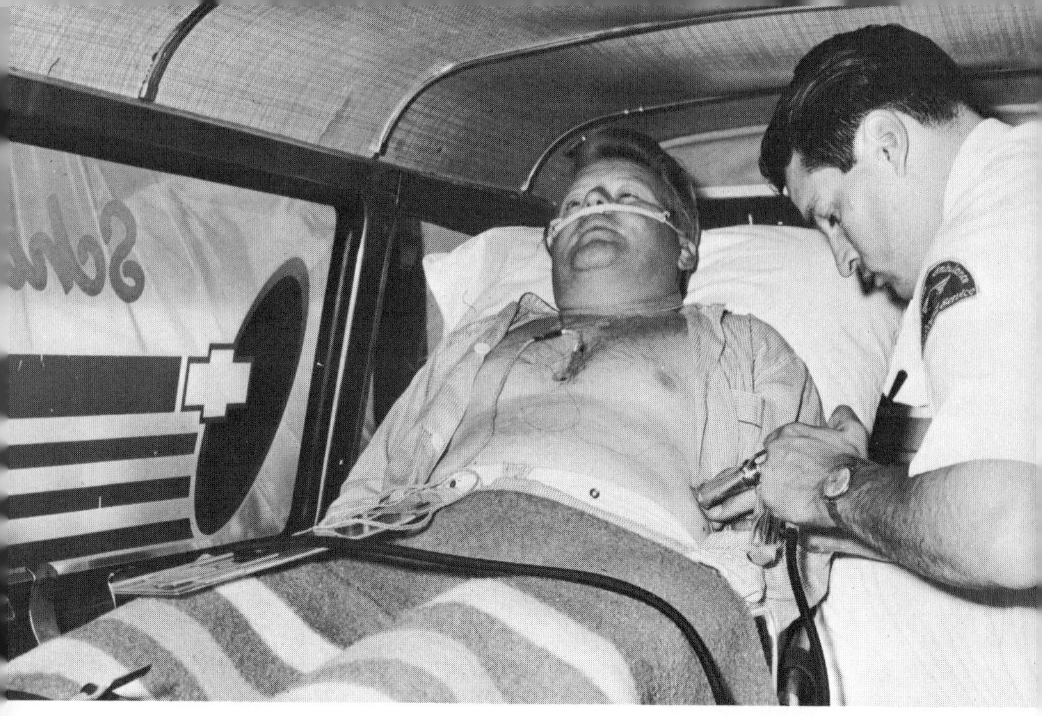

An ambulance attendant needs less than 2 minutes to apply spray-on sensors and attach them to an electrocardiograph, making it possible to transmit the patient's heart signals to the hospital while the ambulance is en route. Spray-on sensors were originally devised for use with astronauts and test pilots. (NASA)

The spray-on electrodes are also being evaluated by the Trauma Research Group of the University of California at Los Angeles, which is studying emergency medical systems, and the University of Southern California's Hollywood Presbyterian Center for the Critically Ill. The dry electrodes prove to be particularly successful on children, who are much more active than adults.

Another medical dividend from space bioinstrumentation is the impedance plethysmograph. This device measures variations in the blood supply and requires only four plastic-tape electrodes, or pick-ups: two about the patient's neck, one about his chest, and the other about his stomach. Thus there is now available a simple, bloodless method for measuring the ventricular stroke volume, or volume of blood pumped by the heart per unit of time. The machine is automatic and causes a minimum of discomfort to the patient. It has been used with success by Dr. Denton Cooley, the renowned heart surgeon at St. Luke's Hospital in Houston, to measure the performance of his heart transplants in patients following their operations.

The miniature mass spectrometer, developed jointly by NASA's Flight Research Center and Consolidated Systems Corporation of Pomona, California, also has applications in the hospital and in medical research. Designed for use with high-speed aircraft test pilots and astronauts to analyze the gases breathed out by them during missions, the unit weighs only 28 pounds and occupies less than a cubic foot of space. It is capable of monitoring 12 different gases simultaneously. Because of its light weight, it could be easily developed into a portable unit for use by anesthesiologists. Also, it can be used in determining heart output by the indirect Frick method—which makes use, for example, of carbon dioxide rebreathing.

A small sensor designed to study the mechanical stresses inside solid propellants for rockets finds use in helping medical researchers study the elasticity of bones in living people. Such studies ultimately may help to show why bones grow brittle with age and help to provide means for preventing such conditions. Sensors that detect "hot spots" in the human body by measuring the infrared radiation from it owe much of their technology to that used in developing horizon sensors for spacecraft. These "hot spots" are indications of diseases or other abnormalities. A special space sensor used to determine the viscosity of solid propellant slurries for rockets appears useful in measuring the viscosity of the blood. The flow of blood in tiny capillaries can also be measured by a sensor developed originally to measure the vibration of spacecraft.

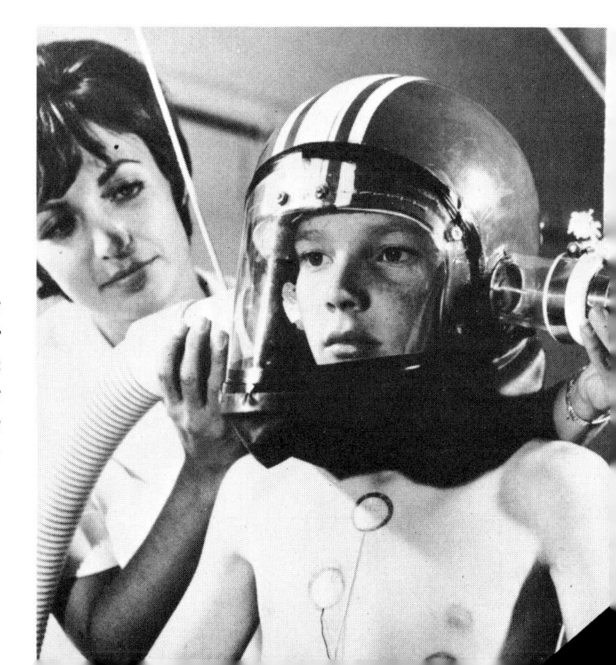

Headgear worn by astronauts has been adapted as a respirometer, for measuring oxygen consumption while the subject exercises. Air is drawn in one side and goes out the other to a breath analyzer. (NASA)

A heart monitoring system for several patients at a time is also the result of space research and bioinstrumentation. The Hamilton Standard Division of United Aircraft Corporation, Windsor Locks, Connecticut, makes one that applies space-age telemetry to the needs of the modern hospital.

The RKG Monitor System can simultaneously record the electrocardiograms of four patients, allowing one nurse to serve patients in rooms up to 250 feet away from her console. Since the sensors on the patients require no wires—the data are transmitted by tiny FM radios which can be carried in the pocket of a bathrobe—the system can be used with patients sitting up or walking around in their rooms.

The receiving console has an oscilloscope which displays at the same time the ECG traces of all four patients. Its individual heart-pulse meters have audiovisual alarms, which are triggered if the heartbeat exceeds low or high limits set by the doctor. A special rate meter with a built-in alarm automatically turns on the recorder to make a permanent chart of abnormal heart rates. The RKG System is now in use in over 400 hospitals in the United States and Europe.

Space instrumentation may also find a useful role in diagnosing and treating the mentally ill. A brain wave sensor and radio transmitter developed at NASA's Ames Research Center for use with high-speed test pilots and astronauts has been tested for such purposes at the Agnews State Hospital, San Jose, California. Since the device does not require electrodes to be placed under the scalp or the shaving of patches of the patient's head, it is much more comfortable and the patient remains calmer. Also, since it operates by radio, there are no wires connecting the patient to a machine. This feature is important because many patients identify wiring connections with shock treatment and become anxious, either from having undergone such treatment or because they fear to experience it.

The device's electrode is able to detect brain waves through the hair and scalp, and the radio transmits them directly into a computer for analysis. By observing the patient's brain wave response to flashing lights, doctors can distinguish between various types of schizophrenia and prescribe appropriate drug therapy. The transmitter for the unit is only 0.5 cubic inch—little larger than a cube of sugar. Doctors at the Agnews Hospital feel that the instrument has a potential in other areas of mental health research as well. Patients could

A child plays while her brain wave patterns are recorded by sensors in her soft cap, transmitted by radio to a telephone, and monitored in a laboratory miles away. The cap, developed for use in space exploration, permits complete freedom of movement within a radius of 100 feet of a telephone receiver. (BRAIN RESEARCH INSTITUTE, THE CENTER FOR HEALTH SCIENCES, UNIVERSITY OF CALIFORNIA AT LOS ANGELES)

wear it in their wards as they engaged in a variety of activities or were presented with different kinds of situations, and data on their corresponding mental state could be radioed to a computer.

Veterinary medicine also finds applications for telemetry systems developed for astronauts. Dr. G. Frederick Fregin, a New Jersey veterinarian, adapted space bioinstrumentation for his basic research into the heart action of horses. He used equipment developed by the Electronics Components Division of United Aircraft Corporation, after trying several other types of instruments. Dr. Fregin's experiments indicated that a racehorse's heart rate climbs as high as 260 beats per minute when it is "in the clubhouse turn." Such knowledge is helpful to veterinarians in understanding the metabolism of animals.

Space research also points the way to future developments in biotelemetry. These include instruments that require no sensors attached to the body as well as very small sensors that can be implanted in the body or simply swallowed by a patient.

Typical of the first of these is the field effects monitor, invented by Dr. William A. Shafer, Convair Division of General Dynamics Corporation in San Diego. This instrument can detect and measure the pulse and breathing rate by receiving and amplifying the signals generated by the blood flowing through the circulatory system. The antenna for it can be built into a chair or mattress, and measurements can be made while the patient is sleeping or resting. It has great promise for monitoring patients who must be confined for long periods to their beds.

Other sensors can be incorporated into clothing. Scientists of Lockheed Missiles and Space Company, Sunnyvale, California, have developed such a device that can accurately count the heartbeat of an astronaut laboring under great strain. The cardiotachometer, as it is called, provides information on heart rate as the astronaut moves about on the lunar surface or inside a space station. Its use in the lining of a "gray flannel" suit on Madison Avenue, sometime in the future, might prove of even greater value to heart specialists!

Bioinstruments for use inside the body are called endoradiosondes. No larger than a "24-hour" cold capsule, they can be swallowed or implanted by a surgeon and continuously transmit information on temperature, heartbeat, pressures, and stomach acidity. Endoradiosondes are self-powered by small batteries that can be recharged by beaming radio energy through the walls of the body to them. Current research in this field of space bioinstrumentation is looking at means of using the electricity generated by the body itself to power such instruments.

The future will also see communications satellites relaying data from bioinstrumentation. The monitoring of the hearts of patients in one country by medical experts in another was demonstrated on 3 July 1967, when an electrocardiogram, vectocardiogram, and other physiological data of a patient in Tours, France, were sent to the U.S. Public Health Service in Washington, D.C., via the Early Bird communications satellite. Doctors in Washington examined the ECG trace as it was displayed on a strip recorder and sent a diagnosis back to France within 30 seconds. For the world's remote and emerging countries lacking in modern medical facilities, this technique offers great possibilities. The Intelsat II communications satellite Pacific 1 was used for a similar experiment 9–10 April 1968, with signals being sent from Tokyo to Houston to Washington and back to Tokyo.

OTHER NEW INSTRUMENTS FOR THE DOCTOR AND HOSPITAL

In addition to the bioinstruments discussed above, a variety of other instruments for the doctor and hospital have accrued through space research and technology. Many of these can be adopted directly, while others require some degree of modification.

A case in point is the use of a "walking chair" originally developed as a means of locomotion for astronauts exploring the Moon. The vehicle has six legs rather than wheels. The idea behind the design is that such a vehicle could traverse the rocky surface of the Moon

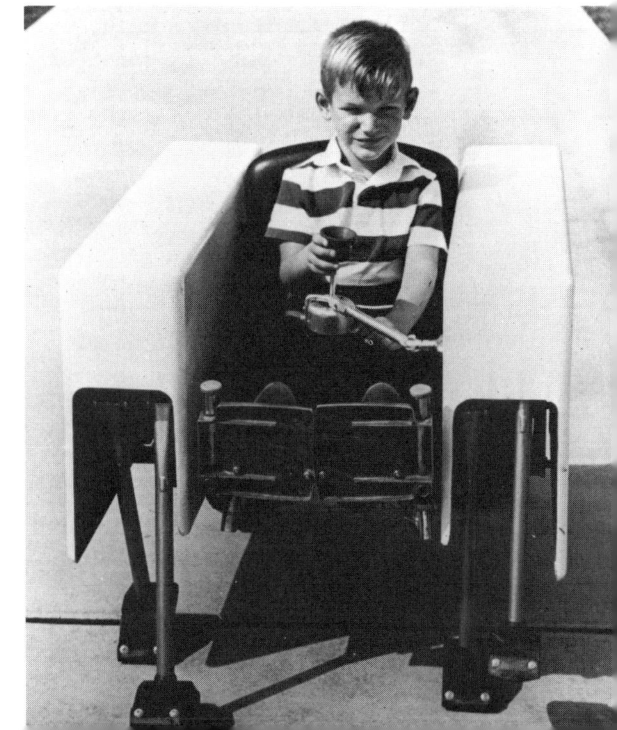

Originally developed as a means of locomotion on the rough surface of the Moon, the lunar walker has been adapted for use by disabled persons. It can climb stairs, walk on beaches, and traverse many obstacles that would block ordinary wheelchairs. (NASA)

better than one with wheels. A small model of it proved just the right size to hold a crippled child. With only minor modifications to the control system, it was found that a disabled child could quickly learn to operate the vehicle—even crossing streets, stepping up over curbs, and strolling on sandy beaches, where conventional wheelchairs cannot go readily. The walking chair can be equipped with different types of controls accommodating young patients variously incapacitated. The feasibility of a chin-strap control and of a single airplane-type joystick has been demonstrated.

The physically handicapped can also benefit from another dividend of space technology, the sight switch. Originally developed as a means of permitting the astronaut to perform certain switching tasks without using his hands, the unit can now be adapted to a variety of electrical appliances, as well as to the controls of a wheelchair, allowing the paraplegic patient to do more things for himself, particularly at home. Prototype wheelchairs equipped with the sight switch have been provided by NASA's Marshall Space Flight Center to the Texas Institute of Rehabilitation and Research, in Houston, and the Rancho Los Amigos Hospital, in Downey, California. The two chairs were built by Hayes International Corporation, Huntsville, Alabama, a company that specializes in aircraft and spacecraft construction. They will be evaluated by the institutions and tested for further refinements and development as a new prosthetic aid for the paralyzed.

The switch is built into the frame of a pair of glasses. Very small infrared light bulbs are mounted on it, projecting a beam onto the white of the eye. As long as the beam falls on the white portion, most of it is reflected back to the unit. However, if the eye is moved until the darker iris cuts into the beam, more energy is absorbed and the change can be noted electronically. Normal blinking of the eye does not affect the unit. By coding glances left and right through an analyzer and system of relays, the paraplegic can cause his wheelchair to start, stop, turn left, and turn right. He also can ring a bell for assistance; turn on and off radios, television sets, fans, and electric lights; or he can open and close doors and windows. It is even possible for him to use a specially modified typewriter. In addition to its use by the handicapped, the sight switch could be applied in industry by workers who "need another pair of hands."

Although it may not appear to be an "instrument" in the usual

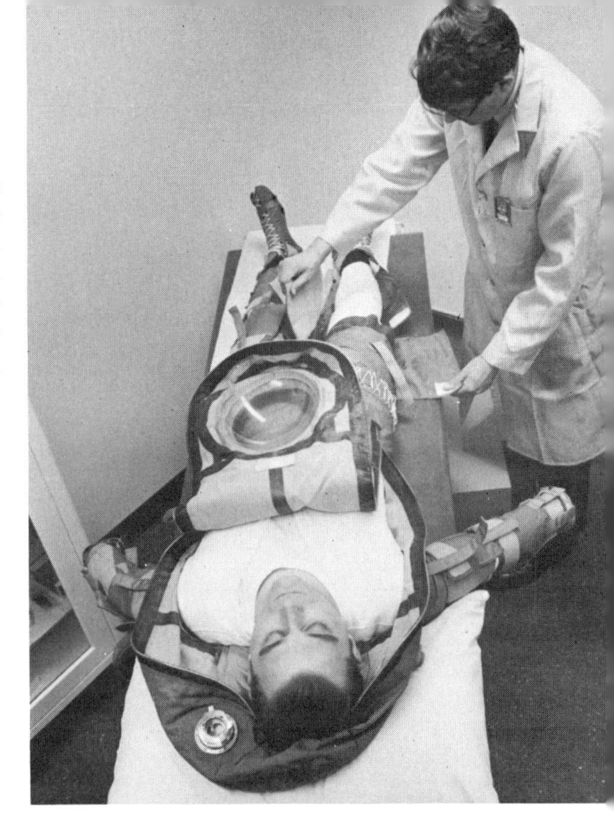

An offshoot of the G-suit worn by astronauts, this suit has multiple cuffs on the arms and legs that inflate and deflate in sequence, assisting victims of coronary attacks by reducing the workload on the heart. (HAMILTON STANDARD DIV., UNITED AIRCRAFT CORP.)

sense, a test pilot's pressurized suit has been used in hospitals to stop uncontrollable bleeding in an emergency. For example, in September 1969 Mrs. Mary Phillips, a housewife and mother of two children, was in the Stanford University Hospital bleeding uncontrollably after a minor operation. She received 46 pints of whole blood and 64 units of plasma in five weeks while doctors sought to control her hemorrhaging. One of the doctors recalled that a pressure suit had once been used to control bleeding during brain surgery at the Cleveland Clinic in Ohio. A call to NASA's nearby Ames Research Center produced a "G-suit," modified to fit the small patient. Placed on her and inflated for 10 hours, it brought the bleeding to a halt. Doctors theorized that the suit had reduced the pressure differential between the blood in the arteries and the tissues outside, allowing normal coagulation to take place. The G-suit is used by high-performance aircraft pilots to keep the blood from draining into the lower body from the brain during centrifugal turns. Thus it prevents blackout in such critical maneuvers.

Drawing on its experience in making spacesuits for astronauts and life support systems for spacecraft, United Aircraft Corporation's Hamilton Standard Division built a heart assistance device for the National Heart Institute. It is a prototype of a unit that could be used in hospital emergency rooms and ambulances to provide immediate medical aid to heart attack victims. Resembling a spacesuit, it can be placed on a prostrate human being in less than 5 minutes. Unlike other forms of heart assistance device, the suit does not require surgery to connect elements of it to the circulatory system.

The suit is designed to do two things: reduce the work load of the damaged heart by lowering the afterload of the left ventricle (which pumps the blood out of the heart); and increase the pressure in the aorta by helping to fill coronary arteries. It functions through the use of multiple inflatable cuffs on the arms and legs. They are sequentially inflated and deflated as the heart beats. An electromechanical pump, triggered by electrodes attached to the patient's chest or arms, synchronizes the cuff actions to the heart. Pressure is applied when the heart is at rest, and is released when the heart pumps (beats), resulting in what is called external counterpulsation. This sequential inflation tends to "milk" the blood backward into the aorta to increase the blood pressure in the central arterial system. With the rapid deflation of the cuffs just before the heart beats, the pressure in the aorta is reduced and the heart does not have to work so hard to pump the blood.

A search for a better means of nondestructive testing and for improved inspection techniques for space hardware, particularly in orbiting space stations, led NASA's Marshall Space Flight Center to support the development by Westinghouse Electric Corporation, in Elmira, New York, of a more efficient, solid state image amplifier for use in radiography. The amplifier panel so produced can store an image for several days, permit photographic prints to be made from it, then be erased. The device can be reused many times. More importantly, the contrast of the image can be varied to enhance details at various depths.

The panel can be employed with a portable source of radioisotopic radiation, in a self-contained unit weighing less than 22 pounds, including shielding. Indeed, the complete outfit can be carried in an attache case quite easily. This portable model should be very attractive to federal and state health departments, particularly for making tuber-

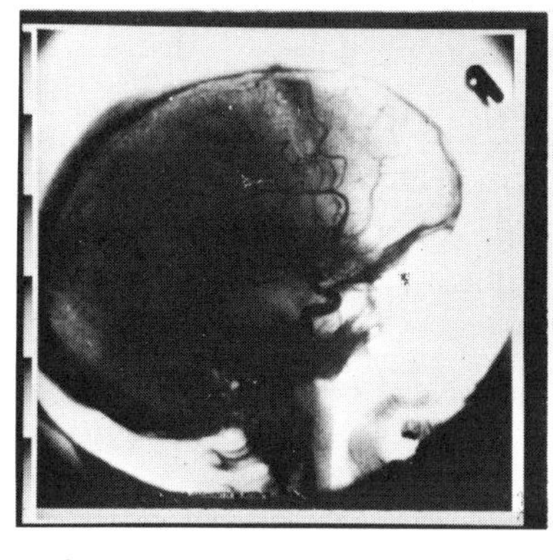

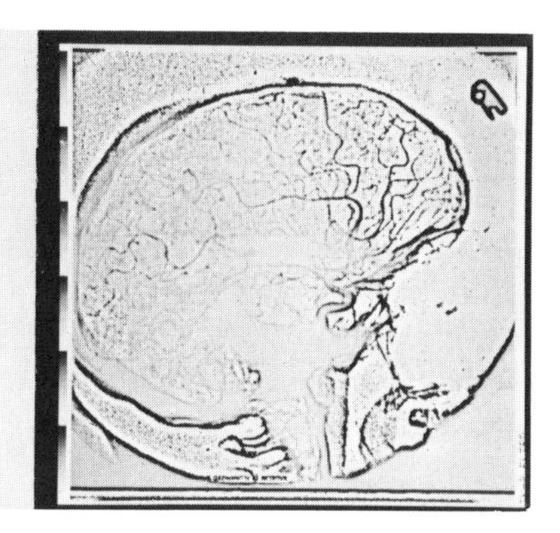

The technique of using computers to enhance the clarity of photographs returned to Earth from unmanned probes of the Moon and Mars also is being used to produce better X-ray pictures. Computer version (right) shows many details not visible in the original X ray at left. (JET PROPULSION LABORATORY, CALIFORNIA INSTITUTE OF TECHNOLOGY)

culosis surveys; and it also has obvious applications for medical teams in disaster relief and for the medical corps of the military, as well as the doctor's office and the hospital emergency room.

Better X-ray pictures are now available to the doctor as a result of a picture enhancement technique developed by Dr. Robert Nathan and Robert H. Selzer at the Jet Propulsion Laboratory (JPL) of the California Institute of Technology. These two scientists made their contribution to medical technology indirectly while coming up with a means of making clearer the photographs returned to Earth by space probes such as Surveyor, Ranger, and Mariner. Their technique consists of an enhancement of frequency and a means of pattern extraction. Frequency enhancement is similar to the control of a high-fidelity radio set when a balance is desired between low and high frequencies. Pattern extraction draws attention to features that might otherwise be overlooked.

In both processes the computer plays a vital role. Given information about frequency distortion caused by the camera, it enhances the picture by cutting down some of its fuzziness. By extracting patterns of electronic interference caused by other equipment in the spacecraft that tend to obscure details, these patterns can be "subtracted" from the original picture as received.

Since the photographic process is adaptable in various situations, the scientists at JPL applied their technique to hospital X rays. Their best results came with X rays of the skull and photographs of the retina of the eye. Many hospitals and medical research institutions express interest in the work being done by JPL. Dr. Leo Rigler, a professor of radiology at the University of California at Los Angeles, says that some radiologists believe that current chest X rays show only one-tenth of what is actually present. "'Photographic methods which would eliminate certain shadows and enhance others would improve our diagnostic capability," he adds.

Likewise, Richard B. Hoover, a physicist at the Marshall Space Flight Center developed an instrument that shows great potential in helping neurosurgeons locate very accurately tumors of the brain. It is being tested by the School of Nuclear Medicine of Vanderbilt University. A collimator for gamma rays, it will ultimately permit doctors to determine within 0.1 mm the location of the tumor. The collimator itself is a tube within which are 15 tungsten plates only 4 inches square, each containing 2,500 holes only 50 mils in diameter and arranged in a random pattern provided by a computer. To utilize the collimator, doctors inject the patient with a radioactive fluid, such as technetium-99, which collects in the tumor and begins giving off gamma rays in all directions. The collimator is placed against the skull and permits only those rays traveling parallel to the optical axis of the device to fall on a photographic plate or a scintillation plate. By vibrating the collimator slightly in a plane perpendicular to the optical axis, the gamma rays "paint" a picture of the tumor that is actual size. If three such space-age pinhole cameras are arranged so that their optical axes are at right angles to one another and intersect in the region of the tumor, three images can be obtained that permit doctors to determine very accurately the size and shape of the tumor in each plane.

A good example of the transfer of technology by adaptation is the vastly improved ballistocardiograph. This diagnostic instrument measures the pumping power of the heart. It is basically a platform, on which the patient rests, which moves in response to his beating heart. Instead of springs, the new machine is mounted on air bearings, which were first used in guided missile and space vehicle guidance systems to virtually eliminate bearing friction in gyroscopes and other precision moving parts. In this bearing, air is used as a lubricant rather than oil or a solid such as graphite.

Since it does rest on a thin film of air, the improved ballistocardiograph is effectively isolated from vibrations that might be transmitted to it through the floor or structure of the building. Thus it does not pick up vibrations from traffic passing in the streets, moving or vibrating equipment in the hospital, or footsteps of personnel in the room. The readings it gives are much more accurate than those obtained from older models.

An especially sensitive ballistocardiograph has been developed by Britain's National Engineering Laboratory and tested by the Royal Maternity Hospital, in Glasgow, in cooperation with the Department of Bioengineering of Strathclyde University. It can detect and measure the forces created by the circulation of blood within the fetus of a pregnant woman. The air-suspended system can be tuned to suppress the effects of the mother's breathing; and a computer, provided with data on the ballistocardiogram of the fetus, separates it from that of the mother.

The ability to thus record the ballistocardiograms of unborn infants could lead to techniques that will permit doctors to save the lives of one-third of the babies that currently die at birth because they are not in good enough shape physically to survive the rigors of normal birth.

The principle of the air bearing also produces a novel and more efficient bed for patients who are badly burned. It is the product of the Institute of Orthopaedics and was developed for National Research and Development Corporation in England. Consisting of a contoured mattress with the shape of the human body, the bed allows the patient to float on a thin film of air. Thus his burns heal more quickly; and he is much more comfortable since his body literally does not touch the bed. In designing the bed, the institute drew heavily on the technology of the ground effects machine, or hovercraft, a technology in which England is the acknowledged leader.

A more mundane but equally useful implement for such facilities as hospitals, fire departments, and police rescue units is a new type of litter or stretcher. It was developed to remove workers injured while working inside the huge propellant tanks of the Saturn 5 space booster. It is adjustable in length for various sizes of person and can also be used to apply traction. Weighing only 20 pounds, it can completely immobilize an injured person and permit him to be withdrawn through an 18-inch-diameter opening. It is already finding application in rescue operations from mines, offshore oil wells, and small craft at sea.

The hospital laboratory technician realizes benefit from space technology also. Aerojet-General Corporation's Space Division, in El Monte, California, produces the Sero-Matic System, an automated process for helping the technician determine the presence of organisms in blood samples that indicate syphilis. The machine takes blood samples and automatically prepares them for microscopic examination. When this work is done manually by the technician, only 40 tests a day can be performed. With use of the machine, however, only 3½ hours of his time is required; and it can produce 200 samples per day.

The laboratory technician benefits from another machine that automates one of his more time-consuming tasks: the detection of bacteria in urine samples. It was originally developed by NASA's Goddard Space Flight Center, Greenbelt, Maryland, to detect possible life on other planets. However, scientists at the center in conjunction with medical experts from the Johns Hopkins Hospital, in Baltimore, modified the instrument to permit quantitative analysis of bacteria in a urine specimen in only 15 minutes–a process that takes days by more conventional assay methods in the laboratory.

Samples suspected of having living bacteria in them are placed in the machine in a lighttight compartment. Chemicals are added to them which cause the bacteria to emit light, which is sensed by a photodetector tube. The brighter the light, the more bacteria present. The assay is possible because the chemical adenosine triphosphate (ATP), which is found in all living matter, glows when another chemical, luciferase (which causes the firefly's tail to light up), is introduced. Special chemicals added to the samples destroy the ATP in any white and red blood cells or tissue that may be present. Thus a positive reading comes only from ATP in bacteria.

Since urinary tract infections are second in frequency only to respiratory infections in the United States, examination of urine is one of the most important and frequently done tests in the hospital laboratory. The Johns Hopkins Hospital alone performs 11,000 such tests each month. Thus the need for a machine that speeds up the time required for such testing is apparent.

Aerospace research also produces other electromechanical devices that have great potential in the hospital and medical research laboratory. Arthur D. Little, Inc., of Cambridge, Massachusetts, developed for NASA's short-lived Electronics Research Center an instrument that can translate human voice sounds into commands for a spacecraft, so that an astronaut can maneuver his craft during the heavy stresses of liftoff and reentry or while his hands and feet are immobilized or restrained.

The machine displays on a small oscilloscope tube the profiles of words as they are spoken. It can also cause single vowel sounds to appear in specific areas of the display. With this capability, Dr. Huseyin Yilmaz, who built the instrument, thought it might have applications in the field of speech research and therapy. Experiments showed that the instrument is helpful in teaching people to speak who have been deaf since birth, and it helps people with severe stuttering problems to speak normally. In addition, it proves valuable in teaching the mentally retarded to speak. A 13-year-old student in such a class was able for the first time to speak three different vowel sounds.

"The machine motivates the subject, defines a target to aim at and provides an undelayed feed-back," says Dr. Yilmaz of his machine. Once its range is extended to handle consonant sounds, a "valuable tool will be at hand for a fresh approach" to speech therapy.

The human heart in several ways benefits from space research. One of the nation's leading centers of research in artificial hearts, the Cleveland Clinic, called upon NASA to help it solve the extremely complex engineering task of artificial heart control. Dr. Wilhelm Kolff specifically requested assistance from the space engineers at NASA's Lewis Research Center, also in Cleveland. Together they developed three successively better systems that duplicated rather closely the natural heart's variable pulse rate and waveforms. One of these devices has worked successfully in a calf, but much more engineering is needed before a reliable artificial human heart can be produced.

The same NASA center has also worked with doctors at Cleveland's St. Vincent Charity Hospital on other aspects of artificial hearts. A NASA engineer, who was himself a heart patient, learned of trouble that doctors were having in measuring and controlling fluid flow and pressure. As a result, medical and space technology combined forces to produce a heart assistance pump that can take over some of the work of an ailing heart, permitting it to heal itself naturally under a reduced work load. The unit synchronizes itself to the patient's heartbeat through sensors on the skin that detect voltages as weak as 1 mv.

The Hamilton Standard Division of United Aircraft Corporation, using space technology, has produced a similar device for emergency treatment of heart attack victims. It is quickly and easily connected to the patient by minor surgery through the femoral arteries in the legs. It proved its ability to save human life in 1967 at the Jewish General Hospital in Montreal when all other therapy failed on a heart attack victim. The same company is experimenting with a more advanced machine called a copulsation pump. Computer-controlled, it assists the weak heart by imitating the healthy heart's pumping pattern. The piston-driven pump consists of two liquid-operated diaphragms mounted in a plastic housing. It is connected to the patient through the aorta.

A more efficient artificial kidney machine should emerge from research being done at Aerojet-General Corporation's Von Karman Center, in Azusa, California, for the National Institutes of Health. Earlier work that the center had done in lightweight water-purification devices led to a reverse-osmosis machine. Scientists at the center felt that the machine could be adapted to purifying the blood of patients with kidney failure. Comparable machines in use today are very large and costly and are found in the larger hospitals only. The development of a portable unit would greatly increase the number of patients who could be treated at home and thus decrease crowding in hospitals. A similar machine, developed in England by Hawker Siddeley Dynamics and introduced in 1967, reduced operating costs from $36 to $7.20 per patient per day.

Components developed for spacecraft life support systems have been adapted by Lockheed Aircraft Corporation and the University of Minnesota Hospital to a means of providing and measuring the amount of pure oxygen that newborn babies use and need. Called an oxygen concentrator, the unit automatically controls the temperature, pressure, humidity, and volume of oxygen in the baby's incubator. Dr. Rolf R. Engel, of the University of Minnesota Hospital, explains:

"We need an accurate measurement of oxygen consumption to confirm our observations that newborns consume more oxygen than we could predict from their carbon dioxide production and heat production." Using the space-age unit, medical researchers at the hospital hope to find out how the use of oxygen is related to the baby's growth rate and general well-being.

The hospital operating room and the medical student receive a dividend in the form of an ultrasmall television camera, also a product of NASA's Marshall Space Flight Center in cooperation with Teledyne Systems Corporation of Los Angeles. It was developed to aid engineers in observing the separation of the various stages of the Saturn 5 space booster during flight, the behavior of its liquid propellants in their tanks during flight, and the firing of engines in test stands. The model weighs less than 20 ounces and is only slightly larger than a pack of king-size cigarettes. Attached to the head of the surgeon as he operates, it transmits a close-up view of the procedure by closed-circuit television throughout the hospital.

An even smaller camera is possible because of advanced research into solid state imaging means developed by the same NASA center and Westinghouse Electric Corporation's Defense and Space Center, Aerospace Division, in Baltimore. Instead of the vidicon tube used in conventional television cameras, the solid state imaging device consists of an array of 200,000 extremely tiny phototransistors. The camera lens focuses the image upon the array, and each of these 200,000 solid state photosensitive elements generates as electrical signal in proportion to the light falling upon it. Processed through amplifiers, they present a picture on a picture tube. Since there is no vidicon tube, it will ultimately be possible to cram the solid state imaging circuits into a package of some 2 cubic inches, using the manufacturing techniques of microcircuitry developed by the aerospace electronics industry.

Using the smaller camera with a miniature lens on the end of a fiberglass cable, a doctor could observe ulcers directly or watch the functioning of heart valves. Light for the camera could be provided from a microlamp—another space product, developed to illuminate panel dials in spacecraft. These tiny bulbs can easily pass through the eye of a needle, yet they give a brilliant light.

Medical education has a boon in the form of an "artificial patient."

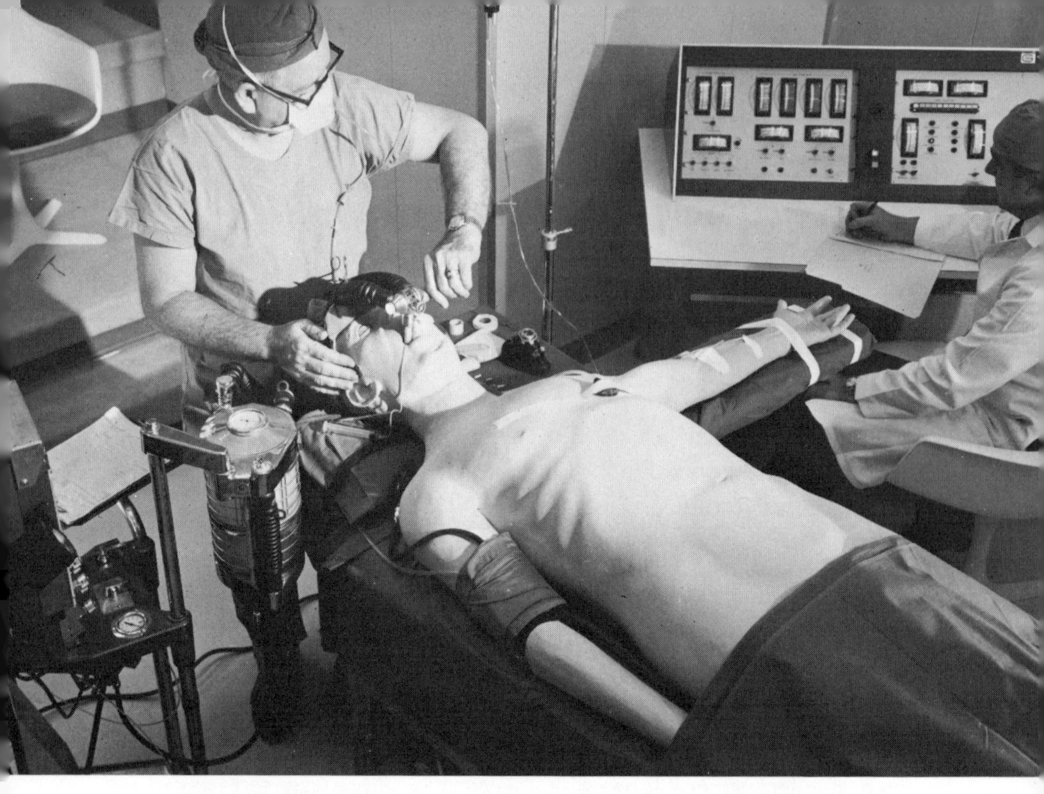

Sim, a lifelike computer-controlled mannikin, has been used since 1967 to train resident physicians in anesthesiology at the University of Southern California's School of Medicine. The sensors, electronics, and other mechanisms inside the simulator cause it to respond in the same way as an actual patient to the student's actions. (AEROJET-GENERAL CORP.)

Sim 1 is a 5-foot 2-inch, 195-pound, computer-controlled lifelike manikin. He is the product of the Von Karman Center of Aerojet-General Corporation and medical researchers from the University of Southern California's School of Medicine. Sim has been used to train resident physicians in anesthesiology since 1967, particularly in the technique of endotracheal intubation—the placement of tubes into the windpipe of the patient. Inside Sim's skin-colored vinyl hide are the sensors, electronics, and mechanisms that simulate movement of the jaws, heart pulse, eye pupil dilation, and wrinkling of the brow—all controlled by the computer in response to the actions taken by the student.

Not only can the anesthesiologist hear Sim's heart beating in his stethoscope and monitor his blood pressure, but he must contend with the possibility that Sim will attempt to spit out the tube and begin retching! Indeed, Sim behaves like a proper patient only when the doctor administers the appropriate amounts of four different drugs.

The instructor, sitting at a console, monitors the progress of the anesthetic process and can at any time interject emergencies such as cardiac arrest, fibrillation of the heart, or blocking of a bronchial branch. He also follows the flow rates of all gases used as well as the amounts of drugs injected. Furthermore, by glancing at his display he can tell whether a face mask is properly fixed, or the position of the air tube in the windpipe. Modifications to Sim include a new right arm and a special-purpose computer that will extend his training capabilities to cover ward attendants, medical students, nurses, and interns. The new arm will have identifiable muscles, nerves, bones, and a pulse. In addition, there will be blood vessels from which samples can be taken. Modified Sim will undergo a two-year program of research, development, and training studies to determine his feasibility and the cost effectiveness of training additional hospital and medical personnel.

Such simulators are not merely expensive exercises in engineering ingenuity. They serve a definite purpose in medical education and materially speed the training of doctors and dentists. For example, they provide "instant" patients, making the scheduling of classes much easier. They can be programmed to display the symptoms desired by the instructor and to do so on command. Also, they are, so to speak, immortal, and as such are able to survive mistakes.

Dentistry receives a dividend in the form of an instrument that will detect teeth areas where decay is almost certain to start long before it can be spotted visually or by other means. The instrument was originally developed to analyze samples of the lunar soil brought back by the Apollo astronauts. Modified for use by dentists, the ionizing radiation detector will be put on trial at the Zoller Dental Clinic of the University of Chicago. Early detection of potential decay conditions will permit preventive treatment with fluorides and save the cost of fillings later.

A variety of other dividends accrue to medicine through space research. Some may seem trivial, but it must be kept in mind that they mark an improvement in service or technique that results in greater patient comfort or well-being. In all probability they would not have come about without the impetus of space technology.

Typical of such dividends is the high-speed dental drill, made possible because of very small and accurate ball bearings developed for scientific satellites. Another is the retraining of physically handicapped persons by the use of machines developed to train astronauts to walk and run in the one-sixth gravity of the Moon. Still another is the special aluminized plastic, used in such balloon satellites as the Echo and Explorer, that has been adopted as swaddling for newborn babies, it being much more efficient to keep the babies warm with this swaddling (they retain 80 per cent of their body heat) than to warm the delivery room to a temperature comfortable to the babies. Of course, the swaddling is especially useful in home or emergency deliveries outside the hospital.

Furthermore, metallic alloys such as vitallium, developed for use in space because they have improved friction and wear characteristics and require no lubrication, are being tested for use in artificial bone joints. Meanwhile, the processes for sterilizing space probe components against microbes seem adaptable for the sterilization of protein-containing materials, such as medicines and vaccines, that must enter the human body. Also, the need to make electronic circuits smaller and lighter has led to a technology that permits the hard-of-hearing to have better aids: devices so small that they fit within the ear and require no clumsy arrangements of wires, batteries, and microphones strung about the body. Advanced models, stemming from U.S. Air Force space research, and using closed-loop body electrical currents for power, are possible.

THE SPACE-AGE HOSPITAL

Space research and technology led Dr. Hugh C. MacGuire, a pediatric surgeon of Montgomery, Alabama, to apply those techniques to the field of hospital design and operation. "It became increasingly clear that medicine would have to adopt the technology of the space age if it were to fulfill its ancient mission of caring for the sick on a basis they could afford," he states in explaining the radical departure from conventional hospitals.

The resultant windowless structure is built of aluminum panels filled with insulating plastic foam. The outside surface is enameled

The circular Atomedic Hospital designed by Dr. Hugh C. MacGuire, Montgomery, Alabama, incorporates many of the innovations and techniques of aerospace research. An Atomedic unit served as the emergency medical facility at the New York World's Fair in 1964–65. (MITCHELL R. SHARPE)

and the inside surface is a baked-on vinyl plastic. Neither surface should need refinishing for at least 20 years. Fireproof material is used throughout the hospital. The circular structure is largely cantilever-supported from a center pylon and is, typically, 136 feet in diameter and 15 feet tall at the center. It can be erected on 1 acre of land. Since it is of modular design, various diameters are possible. Cloverleaf clusters of single units can provide a medical center at a cost far less than for conventional buildings.

The circular design saves medical personnel many steps in caring for the patients in the Atomedic Hospital, as Dr. MacGuire calls his conception. The rooms are placed around the perimeter, and the nurse on duty is seated at a console in the center of the hospital where she can see all doors and, if necessary, all patients. The central core of the hospital also contains the operating area for surgery, maternity delivery area, and four consoles. These are for centralized communications services, a computer and data bank, diagnostic and therapeutic equipment and sterilizer, and food preparation.

Each room typically has 170 square feet and will accommodate one or two patients or six bassinets for babies. Adjoining rooms share a modular bathroom similar to those on airliners (which are easy to maintain and replace as a unit). The bathroom is watertight and the patient can shower while seated. It is also equipped with infrared heat lamps to dry it out. Each room has a television camera to monitor patients in bed and the usual oxygen and suction lines.

Situated on the perimeter also are the X-ray room, darkroom, waiting room, solarium, storage rooms, doctors' and nurses' lounges, and the laboratory, as well as the equipment room, which is isolated structurally from the rest of the hospital so that noise and vibration from machinery and equipment are reduced to a minimum.

As radical as the design is, the appointments are even more so. Patients are provided with biotelemetric devices like those described above. Since the units have self-contained FM transmitters sending data directly to the hospital computer, there are no wires to discomfort or impede the patient, who is free to move about his room. The computer constantly monitors such things as pulse rate and breathing rate. Should either go outside of set limits, a warning light and buzzer are activated on the nurse's console so that she can rush to the aid of the patient. The computer also keeps an inventory of all hospital supplies and drugs and tells when additional quantities should be ordered. As if this were not enough, it also keeps a running bill for each patient.

Air in the core area of the hospital is treated to several filtering processes to remove solid particles and bacteria. Since the whole core area is so antiseptic, surgery and obstetrics are performed in it without the need for special operating and delivery rooms. The air in this area is also at a slightly higher pressure than in the rest of the hospital, so that dust and airborne bacteria are prevented from entering when the doors are opened. The air in each patient's room is exhausted to the outside atmosphere to prevent cross contamination by airborne germs.

The Atomedic Hospital was used as the emergency medical facility for the New York World's Fair in 1964–65, and its concept has been refined considerably since then by Health Resources Corporation, Atlanta. This company drew heavily on experience that accrued to Dr. A. Evan Boddy, who operates an Atomedic Hospital in nearby Woodstock, Georgia. Although the concept is ideal for small towns, isolated geographical regions, large city ghettos, emerging countries, civil defense contingencies, and the armed forces, a market has failed to materialize for this economical hospital facility. There are several factors that work against it: antiquated state and local building and zoning regulations, apathetic attitudes of medical personnel and members of trade unions, inertia and lethargy of bureaucrats on state hospital boards, and lack of interest among entrepreneurs and financiers generally. Ironically, the greatest drawback to Atomedic's growth is the federal government, which subsidizes a great proportion of the bill for community hospital construction in this country today. The requirements for aid under the Hill-Burton Construction Act are rigidly laid down with conventional hospital architecture in mind.

The Atomedic Hospital is extremely economical. Its economies come not only through standardized achitecture and prefabricated structural components but also through a design that permits a staff-to-patient ratio of 0.9 to 1, as opposed to the 2.3-to-2.8 staff-to-patient ratio required in conventional hospitals. Economies are effected in other ways, too. There is little laundry because there is practically nothing to wash: sheets, pillowcases, towels, drapes, surgeon's gowns, are made from paper and burned after use. This practice also decreases the chance of spreading infection throughout the hospital when collecting used linen on a room-by-room basis. Also, food is prepared by a caterer and delivered to the hospital, where it is served in disposable utensils. Thus there is no need for a kitchen and staff of food handlers.

Aerospace research and technology have also been applied to developing "instant" hospitals for the U.S. Army. Garrett Corporation, which designed and built life support systems for the Gemini spacecraft, has used its skills to devise the MUST (Medical Unit Self-contained Transportable), a 400-bed hospital that can be transported by aircraft to front-line areas, then set up and put into operation within 30 minutes. The various wards and other facilities are inflatable structures blown up by air compressors. They remain inflated even though people go in and out of them through doors or bullets penetrate them.

To accompany MUST, North American Rockwell Corporation's Space Division has provided an air-transportable pharmacy. Capable of operating in temperatures from –65°F to 120°F, it can store and dispense 500 different drugs, including those that require refrigeration. It also has its own water purification plant.

The civil applications of instant hospitals and pharmacies are obvious, too. They are ideal for civil defense and natural-disaster relief. Prepositioned around the country in storage, they could be flown in hours, or less, to areas devastated by atomic warfare or natural disasters such as floods or hurricanes.

In retrospect, it is clear that aerospace research and technology have directly benefited man on the Earth by providing better means of caring for him medically. Yet much more in this area could be done if the transfer of technical knowledge from the space laboratory and factory to the medical laboratory could be expedited. Furthermore, additional benefits could be realized if the medical profession in general were to take more positive steps to examine and evaluate what already exists as potential for medical science and practice, rather than wait for the aerospace industry to take the initiative in such actions. Finally, the dialogue between doctor and space engineer must be encouraged at every level by professional societies and organizations of both disciplines and by the federal government as well.

3
THE SPACE SYSTEMS APPROACH TO PROBLEMS ON EARTH

ONE OF THE LEAST-KNOWN BUT POTENTIALLY greatest benefits accruing from space research does not involve hardware. Rather, it is a "software" dividend. It is a technique for the management of large, complex, and expensive projects that produces the optimum product in the shortest time at the best cost. While the money associated with a trip to the Moon may seem to belie this claim, the results prove otherwise. The taxpayer gets more for his space dollar than he realizes.

The technique is known by several names—operations analysis, operations research, systems engineering, and cost effectiveness, to name a few. For simplicity of discussion, "systems approach" seems most descriptive for this chapter.

One of the best definitions of the systems approach is that of R. A. Johnson, F. E. Kast, and J. E. Rosenzweig in their book *The Theory and Management of Systems:*

> The systems concept is primarily a way of thinking about the job of managing. It provides a framework for visualizing internal and external

factors as an integrated whole. It allows recognition of the function of subsystems, as well as supersystems within which businessmen must operate. This systems concept fosters a way of thinking which, on the one hand, helps to dissolve some of the complexity and, on the other, helps the manager to recognize the nature of complex problems and thereby to operate within the perceived environment. It is important to recognize the integrated nature of specific systems, including the fact that each system has both inputs and outputs and can be viewed as a self-contained unit.

Antecedents of the systems approach in both industry and government are several and varied. In the mid-1930s, R. H. Macy & Company of New York established a Research Division consisting of mathematicians, economists, and statisticians that applied an early form of the systems approach to improving the company's merchandising and administration. Similarly, during the same period, Radio Corporation of America began planning for television broadcasting. Again, during World War II, an emerging systems approach produced radar in England and studies that determined the optimum number of ship convoys under varying circumstances.

However, the refinement and optimization of the systems approach are clearly a product of aerospace management during the formative years of the space age.

HOW DOES THE SYSTEMS APPROACH WORK?

A very good explanation of the sequence of steps in the systems approach is that of Dr. Eberhardt Rechtin, director of the Advanced Research Projects Agency of the U.S. Department of Defense. In the June 1968 issue of *Astronautics & Aeronautics* he writes:

> Systems engineering is a methodology for the solution of complex problems consisting of the following six steps:
>
> (1) Statement of the problem to be solved.
> (2) Establishment of quantitative objectives in order of importance.
> (3) Identification and quantitative description of all significant elements

(subsystems) and their interrelationships (interfaces)—e.g., the important technical, legal, contractual, organizational, social, and political parts of the system and their effects on each other and on the objectives.

(4) Design of the system, including the tradeoffs among the competing characteristics of the system, subsystems, and interfaces.

(5) Detailed design, construction, and test of the system.

(6) Integration, test, and evaluation of the system.

The systems approach has several characteristics that make it attractive for problem-solving outside the field of aerospace today. For one thing, it is in step with the times, being concerned primarily with cost effectiveness and economy. It is most effective when put to work on large, complex problems—of which there are many. Also, it can encompass and handle efficiently a large number of constantly changing factors such as money, time, people, and environmental and social conditions.

The key to a successful systems solution to any problem, technical or sociological, lies in the proper and accurate accomplishment of the first step—stating the problem. Considerable time and effort must be spent in determining just what the problem is. It may not be always possible to state it simply as "eliminate poverty," "build a spaceship capable of reaching Mars," or "provide a stock transfer system for Wall Street." If the problem is understated or overstated, the proposed solutions may fall far short of the objective or be exceedingly wasteful of resources.

The problem also should be put in numerical terms since the systems approach depends heavily upon mathematical description for proposed solutions. By reducing the problem to an equation, the systems analyst makes a "math model" of it. He can change the model by putting in different values for many variables, and then study the results. In the math model of a port operation, for example, he can study the effects produced on the operation when a strike occurs, or when a ship ties up ahead of time or does not appear on schedule. Likewise, the impact of a fire that destroys a dock can be determined. In a similar fashion, the designer of a new airport can predict the effects of a crash on a particular runway, or of a heavy fog on the operation of the field as a whole. The police chief can

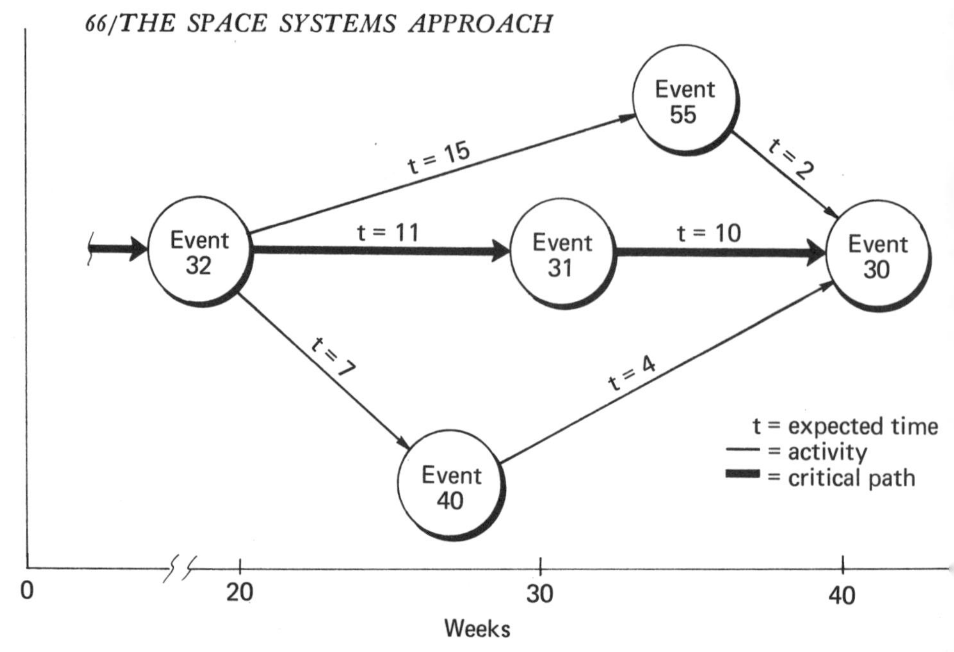

Charts permit managers to tell at a glance the status of a project and how to readjust their resources to keep it on schedule as its various phases change, in the PERT (Program Evaluation Review Technique) management system pioneered by the aerospace industry. (RICHARD P. MCKENNA)

likewise predict the place that crime is most likely to occur in a city —and the time of crime as well.

The future for the systems approach, particularly in the United States, is well expressed by Dr. Simon Ramo in his book *Cure for Chaos: Fresh Solutions to Social Problems Through the Systems Approach:*

> So, perhaps in the end, the systems approach is most essential as a tool to be developed and applied because it encourages and makes possible action. We badly need that encouragement in a society of people who must, in majority, move along together in their thinking, approval, interest, and appreciation before such action is possible. In fact, we are today controlled too much by crisis action; nothing gets done until a problem reaches crisis proportions. Then we are likely to go off in a frenzy. The habit of the use of the systems approach, if we can acquire it, will provide a steady flow of clues to predict and forestall cataclysmic effects of inaction.

By the beginning of 1971, the systems approach was being used in a variety of nonspace applications, particularly those in the social areas.

MEDICINE AND HEALTH

Anyone who has been hospitalized lately can well envy future patients of a hospital being built in the Canadian province of Alberta. The University of Alberta, in 1967, hired a team of systems engineers of TRW Corporation to apply its skills in the design and planning of a new $100-million medical complex. In the first four months of the systems approach, the Canadians realized a savings of $6 million. Over the entire project, savings might run as much as $12 million to $14 million under estimated costs.

How did space engineers go about working on a hospital and medical school? They sat down with hospital staff members to determine what the objectives of the new complex were to be. These were then assigned one of two priorities: "required" or "desired." Among the many things analyzed was the flow of personnel and material between different parts of the school and hospital. The types of trips were examined from the viewpoint of importance and weighted numerically—doctors on the way to a crisis got a high score; visitors to the hospital, a lower one. These were integers that entered into the cost considerations for several architectural designs. Various plans were examined in this light, and the most cost-effective one was selected.

TRW Corporation has also worked with Montefiore Hospital in New York in planning a special operating room equipped with the latest advances in scientific instrumentation and data processing equipment, much of which derives from space technology. Among the other systems studies made by the company is one for the University of Kentucky to design an automated drug laboratory that considerably reduces the errors made by human beings in the preparation and dispensation of drugs in a hospital.

The systems approach has led to a computerized medical information system for the Daneryd Hospital in Stockholm. Eventually it will be the information manager for a group of hospitals in the Swedish capital, storing medical, financial, administrative, and planning information. While an inventory is being computed for one hospital, the name of a patient and his medical history can be displayed on a TV screen for an inquiring doctor at another. Other TV displays will show heartbeat, temperature, and respiration of any patient selected for monitoring by a doctor or nurse in any hospital of the group.

A typical example of the versatility of the systems approach in solving smaller problems in the field of medicine is one provided by the Stanford Research Institute for the Palo Alto–Stanford Hospital Center in California. The center had difficulty in scheduling general staff nurses so as to give the maximum number of nurses the largest possible number of weekends off. The systems team went to work interviewing the preferences of 325 nurses. The answers they received were combined with other factors, such as nursing care needed by each section of the hospital and the rotation of nurses among three shifts. The team came up with three cyclic schedules for days and three for nights. The study produced an improved ratio of weekends off, from the former policy of one weekend in four to a new one of every third weekend off for the day shift and every other weekend off for the evening and night shifts.

The utility of the systems approach in medicine of the future was assessed by Dr. James A. Shannon, director of the National Institutes of Health, in the 23 January 1967 issue of *Technology Week*. Eventually, he predicted, engineer-biologist teams would produce math models for studying the operation of the central nervous system and other physiological control systems. Commenting on the relationship between medicine and aerospace techniques, Dr. Shannon wrote:

> I am convinced that biology and engineering are now developing—indeed have developed—a very broad interface. This situation, in turn, is producing close and productive contact between diverse biologists and engineers. Proper exploitation of this relationship through suitable coupling mechanisms will have profound effects on our understanding of disease, our ability to modify the consequences of disease, our understanding of many life processes and their control, our capacity for more precise diagnosis and treatment, our ability to manage our hospitals and, finally, our ability to develop more rational systems of patient care. Yet, the rate at which these objectives can be achieved depends in large measure on the ability of biology to acquire the extraordinary competencies contained in the aerospace industry.

CRIME PREVENTION AND LAW ENFORCEMENT

The annual loss through crime in the United States amounts to some $21 billion. Add to this another $4 billion a year for the operational costs of the nation's criminal justice system, and the total is some $25 billion annually—more than seven times the fiscal 1971 budget of the National Aeronautics and Space Administration. In *one year* crime costs the country more than it spent over a 10-year period to develop the Apollo spacecraft, the Saturn 5 rocket booster, and the overall effort to land the first men on the Moon.

With the appointment of the President's Commission on Law Enforcement in 1966, a first, tentative use of the systems approach to crime prevention and law enforcement was made by a special Task Force on Science and Technology of this body. Charged with looking into the prospects for using science and technology in general in solving the problems of crime, it conducted one of its earliest studies in Los Angeles. Crimes were approached with the objective of apprehension. Not surprisingly, the study showed that the faster the police arrived on the scene of the crime, the more likely it was that the criminal would be apprehended. The group then fashioned a math model of the process that, in effect, showed how to get the best police efficiency in apprehension for the least number of dollars, by looking at alternatives for improving the existing system. The model showed the number of seconds of delay saved, per dollar of operating costs, through:

1. Adding more call boxes in the city so that people could call the police quicker.
2. Placing more clerks in police stations to handle incoming calls for help.
3. Using computers to speed up the process of handling calls coming into the station.
4. Putting more police cars on patrol.
5. Having an electronic police car locator system to allow police dispatchers to send the car nearest the location of a complaint.

On a cost effectiveness basis, the model showed that the computer-assisted communications system was the best of these alternatives. It was several times more effective than adding more police call boxes, for example. The second-best alternative would be to add another clerk per station—but not more than one!

Systems Development Corporation later built such a system for the Los Angeles Police Department, using the IBM Q32 special-purpose computer. Reports on crimes are fed into the computer continuously and are retrieved and displayed in police offices on special closed-circuit TV sets. Ultimately, the system could be tied into an electronic police car locator in such a way that an incoming call for help would be automatically processed by the computer, which would dispatch the car nearest the call, all within a matter of seconds.

In 1965, Space-General Corporation, a subsidiary of Aerojet-General Corporation, undertook a systems study on the prevention and control of crime for the state of California. The systems engineers began with an exhaustive investigation of the state's total process of justice and law enforcement, from the police precinct station to the state supreme court. Among the more interesting findings was the fact that the only individuals in the state who knew the system as a whole were the criminals who had encountered it! The math model

Futuristic police car, called the "Peacekeeper," an outgrowth of Aerojet-General Corporation's work with the El Monte, California, Police Department, features radar and night-viewing television cameras. (AEROJET-GENERAL CORP.)

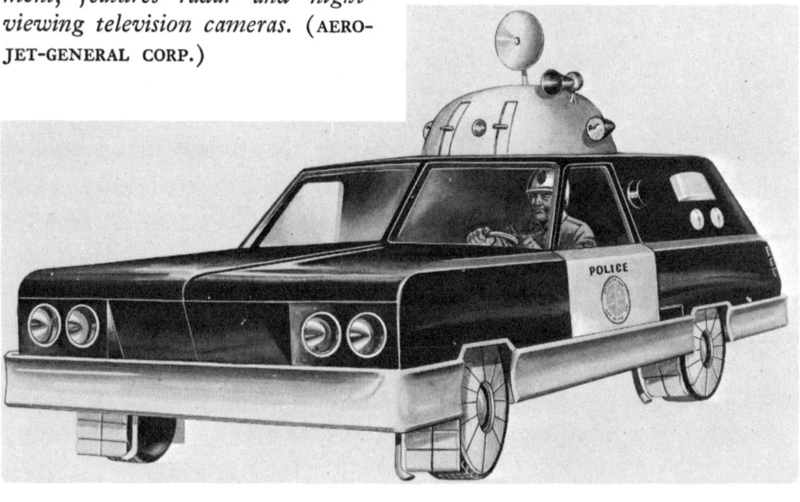

employed proved that a recent sharp rise in crime was not attributable to a breakdown in public morality but simply the result of a great increase in the number of 14- to 29-year-old people, the age group that figures most prominently in crime as a whole.

A more ominous note appeared in another part of the study. A map was made of Los Angeles with various urban areas plotted by such social criteria as maximum population density, low income, a black population of 75 per cent or more, maximum school dropout rate, maximum crime rate. All areas overlapped in the Watts district—which erupted in flames in the summer of 1965, shortly after the study pinpointed it as a part of the city most likely to produce violence.

The study also made significant discoveries about the economics of law enforcement: California spends an average of only $5,800 to punish a murderer but expends $16,900 to deal with a forger of checks! This disparity in costs could be rectified, and a savings of $55 million annually for check forgery convictions be effected, if some means were found for developing an unforgeable check. A million or so dollars in such research might pay handsome dividends in the long run.

The Syracuse, New York, police department conducted a year-long (1968–69) experiment in crime control, suggested by a systems analysis of its operations made by General Electric Company's electronics laboratory located in that city. The study revealed much about the division of responsibility in the police department and the actual functions of the department. Noting that operations were generally defensive rather than offensive, the analysts recommended the formation of a special Crime Control Team, with no duties other than the prevention of crime. The new team did not concern itself with traffic control, parking tickets, rescuing animals, or giving first aid. Responsibility for the team was given to its leader, who deployed it as he saw fit during the hours when crimes were most frequent.

After a year of operation in a high-crime beat, the Crime Control Team approach showed some remarkable statistics when compared with beats using conventional police methods. All crime was down 24.5 per cent from the previous year. Major crimes (murder, rape, robbery, aggravated assault) were down 62 per cent. These compared with citywide values of 8.3 per cent and 29 per cent respectively.

URBAN AND REGIONAL PROBLEMS

The systems approach is also applicable to a variety of urban and regional problems that are becoming more pronounced as the population increases generally. Typical of these is an 18-month study of waste disposal made for the city of Fresno, California, by the Environmental Systems Division of Aerojet-General Corporation. One immediate result—predicted by the study—of the new procedures was a drastic reduction in the number of green bottle flies. Aerojet's systems engineers had suggested that a twice-weekly collection of garbage would result in the larvae of the fly being buried in landfill before they could hatch.

The Aerojet team used computers to examine alternative means of collecting and disposing of garbage on the most cost-effective basis. Aerojet's far-ranging study would require a considerable investment of money to bring to realization, but it could result in an 84 per cent reduction in the adverse effects of waste disposal on the city's environment if it were adopted by the year 2000.

In Japan, Tezuka Kosan Company of Tokyo studied that city's monumental garbage problem. Tezuka designed a machine to compress garbage into a rocklike brick that can be covered with asphalt or cement to form hygienic building materials for land reclamation, roads, or even buildings. Perhaps the urban renewal programs of the future will find ghettos replaced with modern buildings built of the garbage of the former tenants.

American efforts at applying the systems approach to garbage disposal have resulted in a potentially profitable new business. New York City has already reached the paradoxical position of buying more and more garbage trucks to haul refuse to incinerators that cannot handle the load. A systems approach to the problem has resulted in a proposal for a $110-million plant to incinerate both solid waste and sewage. In solving the problem of shrinking landfill area, the new system also would produce heat that could be sold to utility companies for the production of electricity, in a region where power shortages are becoming an annual threat.

The federal Housing Act of 1968 directed the Department of Housing and Urban Development to undertake experiments in large housing projects "utilizing new housing technologies." It would seem that the law of the land is indirectly supporting the systems approach to housing and urban renewal. However, despite the funds available, the existence of outdated building codes and the objections of building-trade unions probably will render this section of the law ineffective. Nevertheless research involving the systems approach to mass housing has been conducted in the United States as well as Europe.

For example, a study by General Electric Company's Missiles and Space Division of the problems of producing housing for dependents of the armed forces resulted in a solution that would be equally applicable to building mass housing for expanding urban or suburban areas. GE's engineers proposed a portable manufacturing plant composed of 22 ordinary trailers that could be joined together and could be moved from site to site. Once the land has been cleared, the mobile housing factory moves in and starts work.

The systems approach to the construction of housing is much more advanced in Europe than in the United States. In fact, European firms in England, France, West Germany, and Denmark license firms in America to produce housing with standardized components that are manufactured on or near the development site. The capital required for such agreements is surprisingly low: from $1 million to $2 million.

Typical of the systems approach to large housing projects in urban areas is Thamesmead, in London. Designed for 60,000 people, it consists of a variety of multistory apartments, two-story row houses, and high-rise structures. Despite the mixture of architecture, all walls, ceilings, and floors are manufactured in a nearby factory and delivered to the construction site to be installed by semiskilled labor. The various concrete walls are cast with windows, electrical conduits and outlets, plumbing—all in place and arranged to connect with mating fixtures in ceilings and floors as needed.

Perhaps the most successful demonstration of the systems approach as applied to regional economic problems is that of Litton Industries, Inc., which signed a contract with the government of Greece to perform multiple studies to help that country realize its potential in mining, farming, and tourism. The studies concentrate on the island of Crete and the western Peloponnesus region of the mainland. The Litton effort identifies potentially profitable enterprises, writes pre-investment prospectus documents, and then locates investors for the projects. Under the terms of the 4-year contract, Litton operates on a cost-plus, 11 per cent fixed-fee basis.

Among the approaches taken in mining is a proposal for exploiting a gypsum deposit, which could profitably produce sulfuric acid and cement clinker. In farming, the Litton engineers developed a computer program to analyze the problems inherent in expanding the raising of cattle. Among the several factors the mathematical model and the computer had to consider were the optimum feed mixture for Greek cattle, the amount of land available for cattle, and the country's commercial meat distribution system. The complexity of the problem was increased by the "inefficiency" of Greek tastes. The average Greek likes young, or "baby," beef; thus cattle for local markets must be slaughtered when they weigh some 400 pounds, rather than the 900 pounds common in the United States.

EDUCATION

Seeking to diversify their services as the nation's space program shrank in the late 1960s and early 1970s, some aerospace companies turned to the field of education.

Using the systems approach, Lockheed Missiles and Space Company developed a course entitled "Drug Decision," which during a period of less than half a year was used by some 40 school districts in New York, New Jersey, California, Texas, North Carolina, Alaska, and Canada. During the period more than 50,000 students took the course. Designed for seventh- through ninth-graders, the course is described by Kenneth T. Larkin, director of Lockheed Information Systems, as a "*total* educational service . . . aimed at students, teachers, parents, staff, and community." Basically, the package consists of a teacher orientation program, a parent orientation program, and a classroom unit with student workbook and a series of reinforcing films. A game to demonstrate methods of attack on drug abuse, played by three students at a time, is also a part of the classroom aids.

From the beginning of the course, the student is involved as a decision maker rather than as the passive recipient of a sermon against

Students play the antidrug game devised by Lockheed Educational Systems as part of a 15- to 20-hour course on the dangers of drug abuse. Players take the roles of community mayor, health officer, and narcotics officer. If they work together, they can beat simulated drug pushers on the board; if they don't cooperate, the pushers win and the community loses. (LOCKHEED MISSILES AND SPACE CO.)

drugs. He alternates between watching the movie and making responses in the workbook. He then sits on mock parole boards and listens to case histories to decide on the rehabilitation of drug addicts. Since more than 90 per cent of states require some drug education in the junior high school age group, the sales potential of the course is obvious.

The same company also made an important systems study to solve a problem facing a large segment of the students in San Jose, California, public schools—Mexican-American children handicapped by knowing only Spanish or a minimum of English. The R-3 program (for "readiness, relevance, and reinforcement," in the space-age jargon) was designed after a careful study of the reasons why such students lagged behind their English-speaking peers by as much as two years. Drawing on its experience in training personnel to run satellite tracking stations by using simulators, the company decided on games to emphasize practice in the basic skills of mathematics, reading, and composition.

As the R-3 experiment progressed, school administrators found that they had only 6 disciplinary problems, while in the preceding semester there had been 86 in the same group. An analysis of the project showed that the group was advancing at the rate of 1.3 years per standard school year, compared with the 0.5-year rate that had been anticipated under the former system.

The systems approach also lends itself to more recondite studies. Rita Zemach, of the College of Engineering of Michigan State University, demonstrated the versatility of the approach with a study, *A State-Space Model for Resource Allocation in Higher Education*, for the National Science Foundation in 1967. It resulted in a math model of the university as a dynamic system of personnel, facilities, and equipment that produces degree programs, research, and public or technical services. The university was seen as a group of interacting subsystems or sectors. Each of these various subsystems was analyzed in terms of its inputs and outputs; a math model was then constructed of their interrelationships. The major subsystems were identified as the student sector, physical facilities, nonacademic production, academic production, and administrative control.

The immediate value of the study is that it produces mathematical equations that represent the functionings of the various subsystems as well as of the overall system. By substituting different values in the subsystems equations, the effect on the overall system can be determined. For example, what happens if there is a drastic reduction in the number of freshmen enrolling in a particular semester? Suppose that the state legislature does not appropriate enough money for a new physics laboratory? What can be done at a particular point in the system to maintain a desired total output for the university? Using such a model, the administrators of the university can make alternative plans for any of several contingencies that may face the institution at any point in its operation.

The systems approach serves education in other ways, too. Aerojet-General Corporation has developed for the California Department of Education a computer system of teacher credential evaluation. It significantly cuts the time needed to review all teacher applications received by the state. The program has provided the basis for the state's Total Education Information Systems, available to all educators for processing teacher applications. The same program also assists educators in student programming and accounting procedures.

The U.S. Department of Health, Education, and Welfare (HEW) sponsored a different type of study by Lockeed Aircraft Corporation in San Francisco. Its purpose was to plan an information network on educational research for the Far West Laboratory for Educational Research and Development, one of 20 such regional laboratories sponsored by HEW. This study marked the first expenditure of federal funds to the aerospace industry for studies in the field of educational research and development. Basically, the network is a computer program by which most public schools in northern California and portions of Nevada can draw upon the latest findings in educational research.

TRANSPORTATION

In the year that saw the landing of the first men on the Moon, the United States had some 97 million automobiles. To the frustrated motorist, creeping in rush-hour traffic along the freeways and expressways—designed to speed traffic through the nation's largest cities—it must have seemed that all 96,999,999 other cars were in front of him. For the airplane passenger today, as well, waiting interminably on the end of the runway or circling endlessly over a crowded airport, the speed with which Armstrong, Aldrin, and Collins and the astronauts that followed them journeyed to the Moon must seem truly remarkable.

The systems approach offers hope to both types of passenger largely because it views transportation as a single system, albeit a complex one consisting of many subsystems. It does not see railroads, airlines, inland waterways, ocean lanes, and highways as incurring separate problems to be solved independently, without regard to the consequences of improving one at the expense of others. As Brenton Welling, transportation editor of *Business Week*, has said, "What in the world is the point in having the most sophisticated jet that $21 million will buy, complete with 15 pretty stewardesses, wide seats, movies, a stand-up bar and an upstairs lounge–if you can't get near the airport to board it?"

Perhaps the most widely publicized use of the systems approach to transportation problems is that of the U.S. Department of Transportation's Northeast Corridor Program, established by a presidential executive order in 1964. The Northeast Corridor encompasses Massachusetts, Rhode Island, Connecticut, New York, New Jersey, Pennsylvania, Delaware, Maryland, and northern Virginia. It is a densely populated region, accounting for only 1.4 per cent of the nation's landmass but 20 per cent of its population. The corridor's 94 cities and counties are tightly interconnected, and any change in the existing transportation system of one affects those of the others.

The objectives of the program are:

1. To develop comprehensive regional transportation plans and investment programs for the Northeast Corridor.
2. To evaluate new transportation technology for applicability to the corridor needs.
3. To develop tools and techniques usable for the corridor and for potential future corridors.

These are to be studied in terms of the 1980s and beyond, for passenger as well as freight transportation.

While the Northeast Corridor Program is concerned with studies and analyses rather than the implementation of better means of transportation, the systems approach is being used in related studies for the Office of High Speed Ground Transportation. These were specified in the High Speed Ground Transportation Act of 1965. They include the development of hardware as well as theoretical studies. Again, most of the contractors are primarily in the aerospace field. By mid-1968, some $52 million had been expended by the office on such studies.

The program seeks to develop a demonstration system linking Boston, New York, and Washington. One of the trains produced is the TurboTrain of United Aircraft Corporation. Designed to carry 148 passengers, it has three cars. The TurboTrain, powered by six 550-hp ST6 free-turbine engines, has reached speeds greater than 170 mph between Trenton and New Brunswick, N.J. However, the usual top speed on the run between Boston and New York is restricted to 120 mph, and the average over the 230-mile route is about 75 mph. A similar train, the Metroliner, built by Budd Company, operates over the New York-Washington tracks of the Penn-sylvania Central Railroad. Of a more conventional design using electric motors, the Metroliner nevertheless can attain speeds of 160 mph.

In addition to these demonstration systems, the Department of Transportation is also sponsoring studies for hovercraft and other advanced forms of intercity transportation. Again, these are made largely by the aerospace industry. Typical of them are the Tracked Air Cushion Vehicle concepts of Grumman Aerospace Corporation and General Electric Company's Transportation Systems Division. Hovercraft, as the name implies, travel above tracks rather than on them, riding air cushions which are produced by downward-blowing fans. And the tracks themselves may be of elevated construction, thus minimizing interference with existing traffic systems. Because of the lack of friction, hovercraft can attain very high speeds—as much as 300 mph with present designs, which would use either jet engines or electric motors for propulsion. The Grumman model proposes to use components proven in flight. For example, a "birdproof" windshield for the train would be made of the same material as that in the Gulfstream I aircraft. The seats and restraints would be those

Model of the 300-mph air cushion vehicle developed by Grumman Aerospace Corporation for the Department of Transportation. (GRUMMAN AERO-SPACE CORP.)

from the E-2A Hawkeye, and the power plant would be the JT8D-9 jet engine, used successfully in the Boeing 727 and the Douglas DC-9 aircraft.

A train of this type has been operating for several years in France. The Aerotrain of Bertin & Compagnie routinely travels at 150 mph on a test track, and it has reached 210 mph by using a jet engine with rocket boosters. Significantly for American industry, the U.S. Department of Transportation has been buying test performance and cost data from the French on this unique means of travel.

Studies made by the British firm Tracked Hovercraft Ltd. indicate that such a vehicle could be commercially competitive with first-class British rail traffic in supplying transportation between London and the new airport to be built somewhere outside the metropolitan area. Studies made of a tracked hovercraft include tests on a special track 3 miles long near Cambridge. The track, when extended to 7 or 8 miles, will permit speeds of 250 mph.

The hovercraft has many advantages other than speed. It is silent, its engine does not generate air pollutants, and it has only one moving part. The U.S. Department of Transportation, with the Northeast Corridor in mind, has funded studies with the British company.

Systems engineering has been applied to intracity as well as intercity transportation. The Stanford Research Institute (SRI), under funds supplied from the Urban Mass Transportation Act of 1966, has made a study of the probable transportation systems for large cities in the 1980s. After considering the total urban environment, its transportation needs, and more than 50 alternatives, the systems analysts decided that no one system would suffice. The answer lay in four different "families" of complementary systems, each of which served specific needs. These included means of getting around downtown, getting around in the neighborhood, getting all over town, and getting from town to suburbs and outlying facilities such as airports.

For getting around downtown, the engineers proposed a conveyor traveling at a speed of 15 mph. An alternative suggestion was a fleet of automatic automobiles powered by electric motors, moving in special guideways constructed above the streets. Both systems were designed to haul large numbers of people for relatively short distances at slow speeds. For getting around the neighborhood, SRI's systems engineers suggested a public automobile system for licensed drivers and a radio-dispatched minibus for nondrivers. The battery-powered public automobiles would be stationed at stands along major thoroughfares. The driver would pick up one, drive it to his destination, plug it into a power line to recharge the batteries, and leave it. For trips around town, a citywide network of guideways for automated vehicles would be a solution. These vehicles could move at speeds of 70 mph. On entering the vehicle, the passenger would insert his destination in a computer control system that would automatically route him there. For trips to suburban areas and airports, the mode of travel would be in special vehicles propelled by an electric motor. The guideways for these vehicles would be enclosed tubes evacuated of air so that the automated vehicles could travel safely at speeds up to 250 mph.

Other nations are becoming aware of the potential of the systems approach for solving domestic problems, especially those in transportation. For example, Lockheed has undertaken a study to evolve a long-range plan for the transportation systems of the Sudan. The study was funded jointly by the U.S. Agency for International Development and the Sudanese Ministry of Finance and Economics. It provides a blueprint for the future development of all modes of transportation within the country, including water, rail, highway, and even pipeline. A similar study, limited to international and domestic air routes, was made for the government of Chile by Northrop Corporation's Organization for Development Assistance Programs.

Intracity transportation within the Northeast Corridor also has received the attention of the systems engineer. Westinghouse Electric Corporation has planned an automatic rapid transit system designed for medium-density commuter loads. Known as Skybus, it would provide service every 2 minutes, day and night, for points within large cities and outlying districts. An experimental version of Skybus is now operational near Pittsburgh.

In San Francisco, the $1-billion Bay Area Rapid Transit System was designed from the beginning by systems engineers. It uses automatic trains that can reach speeds of 80 mph and operate only 90 seconds apart. The whole system is run by computer, and the single line will ultimately transport 30,000 passengers an hour. Train attendants will be on board, but only for emergencies and not as engineers or drivers.

That the systems approach in transportation is valuable from a cost effectiveness standpoint is proved by its application in the planning for the Victoria Line in England's railway system. The builders of this new rail line were faced with several policy decisions that had to be made before work was begun. Would it in the long run be more economical to have automatic locomotives rather than those driven by engineers, despite the initially higher cost of the former over the latter? Which would be more economical ultimately, a simple or a complex track layout for Victoria Station in London?

A math model of the proposed new line was examined by computer. It showed that the potential bottleneck would not be Victoria Station—provided that an expensive and complex network of tracks was built to handle the congestion—but Oxford Circus and Finsbury Park Stations. It also showed that automatic trains would in the long run be cheaper to operate than human-controlled ones. Thus the line was built as the systems approach indicated; and the operation of it has been in close agreement with what the computer predicted.

Of more immediate application in transportation needs is a special safety car designed by the systems engineers of the Republic Aviation Division of Fairchild-Hiller Corporation. The car was developed under contract to the New York State Department of Motor Vehicles for $385,000. Using the systems approach, the engineers treated safety of the occupants as a whole rather than dealing with it on the usual gadget-by-gadget basis, as had generally been done in the past. The car will be put into production on a limited basis for extensive testing by the federal government, which will then write performance standards that all automobiles by the later 1970s must adhere to. The design Fairchild-Hiller car could reduce by half the number of fatalities from front-end collisions at speeds less than 50 mph.

Among the several features of the car are doors that lock to the frame under impact, rather than flying open, and that afford protection against side collisions at up to 40 mph; a hydraulically damped bumper that automatically extends an additional 12 inches when the speed of the vehicle reaches 37 mph; and four rigid roll bars to withstand crashes at up to 70 mph. Other safety features include a special wheel control system that has front and rear wheel drive, and power-assisted dual disk brakes. A rear view mirror mounted on the roof provides periscopic vision of all areas behind the car. The fuel tank

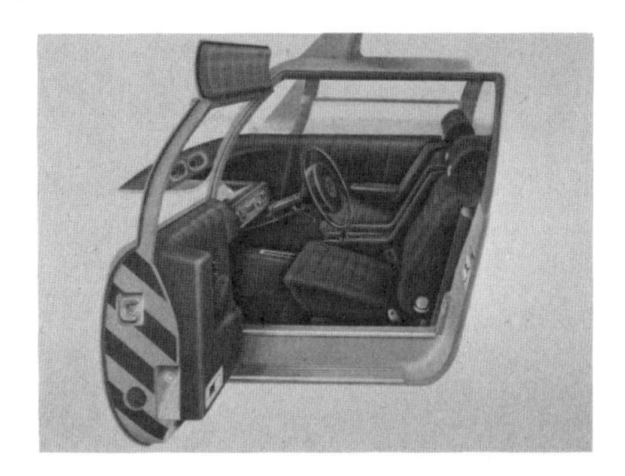

Padded doors to protect occupants against side collisions and a rearward-looking periscope are among features of a safety car designed by aerospace engineers using the systems approach. A prototype model demonstrated in 1967 could have been mass-produced for some $4,000 at the time. (REPUBLIC AVIATION DIV., FAIRCHILD-HILLER CORP.)

is situated in the rear and surrounded by structural members that protect it against rupture from a collision in any direction.

Other systems studies to increase automobile safety include those of TRW, Inc., for the National Highway Safety Bureau of the Department of Transportation, submitted in 1968. This aerospace company applied its systems engineering skills to studying the problems involved in designing automatic checkout instruments for cars. Among the designs projected is a moving rack which takes the car through a checkout center of sensors; these relay performance and other data to a central computer for analysis. Actual construction of such inspection centers may be some time in the future, however. The study included a plaintive note: "At present, the automobile seems to be designed not to be inspected." There are indications, on the other hand, that Detroit is willing to include easily accessible test points in the next generation of its cars.

The transportation of things as well as people benefits from the systems approach. The Industrial Systems Division of Aerojet-General Corporation designed and installed an automated storage warehouse for Union Carbide Corporation in South Charleston, West Virginia. It can handle daily 2,000 chemical drums weighing 500 pounds each. The warehouse features a very high-density storage concept which makes room for 64,200 drums stacked from floor to ceiling. Storage racks are automatically filled from one end by a machine that travels at 75 fpm horizontally and 60 fpm vertically.

Drums are withdrawn from racks by a similar machine. The horizontal and vertical positions of these machines are controlled by a computer. Orders are filled and stocks replenished all automatically. In addition, the system turns over each incoming drum to check for leaks before it is placed in storage.

The Transportation Systems Division of Docutel Corporation has produced a high-speed, automated baggage-handling system for airline terminals. Its lightweight Telecars can carry 100 pounds of baggage, which can be fed into the system at a rate of 1,800 pieces per hour from airline check-in counters. The Telecars are powered by electric motors and travel at speeds of 15 mph on a track that is computer controlled. To operate the system, the passenger places his bags in a Telecar and hands his claim check to the airline agent, who inserts it into a console and adds the flight number to information read from the claim check by the computer. This information simultaneously goes into a memory unit on the Telecar and to the central control computer. The central computer then adds relevant routing and switching information, and off the Telecar goes to the proper baggage-loading area. If the computer is told that the baggage is checked in early, it sends the Telecar to a central holding area until the plane is ready for loading and then automatically dispatches the Telecar to its destination.

ENVIRONMENTAL POLLUTION CONTROL

One of the most difficult problems to which the systems approach has been put is that of pollution of the natural environment. While it may produce valid solutions, in many cases the costs involved would be prohibitive. The Cornell Aeronautical Laboratory is one of several aerospace research organizations that work in the field. Typical of its activities is a contract with the New York State Conservation Department to study ways of monitoring water quality over the length of an entire river.

Using sophisticated math models and aerial photography, the laboratory found that the Niagara River daily receives pollutants from 84 sources, including 32 tons of oil and grease, 1,000 pounds of cyanide,

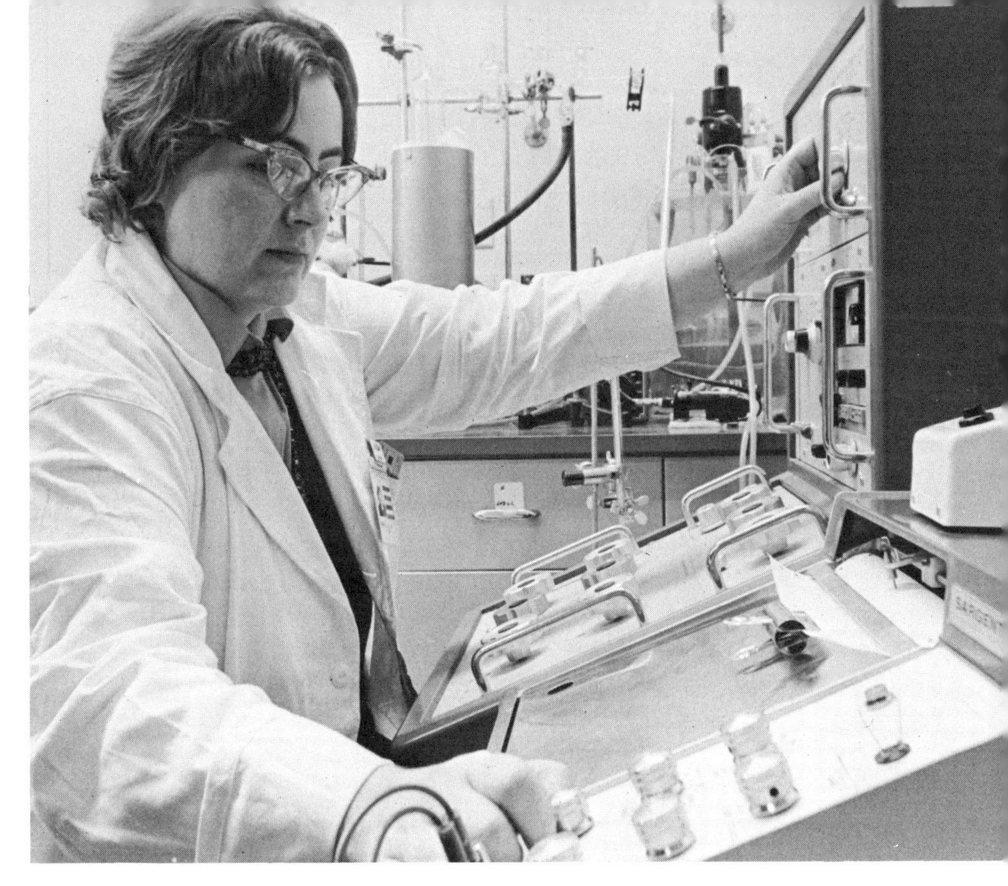

The techniques of systems analysis are being applied to attack complex problems of environmental pollution caused by waste discharges from many sources over wide areas. Here, a technician operates a precision calorimetry system in a study of impurities in water for the Department of the Interior. (ROCKETDYNE DIV., NORTH AMERICAN ROCKWELL CORP.)

4,500 pounds of sulfur, and 1,300 pounds of phenols. The study also revealed that the main causes of pollution were the waters of the already polluted Great Lakes as well as the cities and industries along the river itself.

North American Rockwell Corporation, under contract to the Federal Water Pollution Control Administration, has performed systems analyses of the problems involved in ridding bodies of water of algae by depriving them of nutrition. The company is also studying the use of sensors in detecting air pollutants—sensors that were originally designed to detect spillage of invisible rocket propellants—and has performed studies of problems associated with making fresh water from seawater using equipment earlier developed for space technology.

MISCELLANEOUS APPLICATIONS

The systems approach has been applied also to a variety of other nonspace problems. Typical of these is the call upon aerospace industries to assist stockbrokers inundated with paperwork because of the upsurge of activity in buying and selling of stocks during the 1960s. Some 40 million stock transactions are made daily on the major exchanges. Since as many as 100 people can be involved in each transaction, sometimes almost 4 billion shuffles of paper must be made during a day.

Rand Corporation was given a contract by the two major New York exchanges and the National Association of Securities Dealers to perform a long-range study of the trading system—to find out what sort of system will be needed between 1975 and 1985. This aerospace company used computer simulation and "Monte Carlo" techniques (mathematical approaches originally developed to study gambling odds) to perform the study. The American Stock Exchange also asked North American Rockwell Corporation to study its transfer operations and come up with solutions that could be implemented within two years or less.

The North American Rockwell study produced some recommendations that were revolutionary in a business whose procedures are encrusted with inefficiency. For one thing, aerospace analysts recommended doing away with stock certificates altogether, as owners do not really need to see them or possess them. As might be expected, the study also recommended greater reliance on computers, already available in the big exchanges and larger brokerage houses. It suggested a national clearing service, regional clearing services, and transfer agent depositories. The actual stock certificates would be kept in the depositories, while the daily changing records of ownership would be kept by the clearing services.

Both the New York and American stock exchanges are expected to install automated systems that are generally in agreement with these studies, at a cost of $10 million over a five-year period.

Lockheed Aircraft Corporation has developed a special competence in designing information processing and handling systems for state governments. In so doing, it uses the skills accrued in problem-solving techniques developed during a decade or more of work on complex missile and space systems. States contracting with the company for services in this area by 1969 were California, West Virginia, and Alaska. The objectives of such studies are generally to streamline the information processing machinery of the state governments, to reduce redundancy, and to decrease the need for human clerical components in the system. Generally, the solutions involve the use of computers linked to a statewide communications system with inputs from all state agencies. The potential for improved systems of this sort becomes increasingly attractive to state governments that find it difficult to impose new taxes or raise tax rates.

Information processing systems also find a wide variety of other applications. Typical of these is the system devised for the state of California for issuing and controlling some 700,000 state licenses granted each year to a variety of professions that include doctors, lawyers, undertakers, accountants, and beauty shop operators. The aerospace industry has also assisted the same state in drawing up more efficient plans for emergencies occasioned by natural disaster or civil unrest. These include an information processing system that keeps constant tabs on the available amounts and types of blood in the state's many hospitals and blood banks.

The potential market for programs of information management is so profitable that North American Rockwell Corporation formed a separate company in 1969 to provide such a service, using the experience it had gained in developing and handling complex information and documentation systems for the Minuteman missile, Apollo spacecraft, and the second stage of the Saturn 5 rocket. As early as 1966, the parent company received a contract to provide an information processing and control system for the 21-dam Bonneville Power Authority. Later contracts provided the U.S. Army Materiel Command with a system to improve procurement operations, and the state of California with detail procedures for moving water more than 400 miles from northern to southern California. The corporation also has been active in devising similar information systems for law enforcement agencies, education departments, air traffic control, and environmental pollution control.

HOW APPLICABLE IS THE SYSTEMS APPROACH?

Following the early success of the systems approach in the aerospace industry in the 1950s and 1960s, some managers touted it as a universal answer to all problems. The same approach could as well be used, they said, to solve the really big problems of the world—overpopulation, famine, pollution of the environment, inefficient use of available natural resources, and the like. The idea is simple enough in theory: isolate the problem and examine the alternative solutions, then proceed in a systematic way toward established objectives.

However, a fundamental point is almost always overlooked by the space enthusiasts of the systems approach for nonspace problems. They are used to working within a clearly established customer-contractor framework. The customer is always clearly defined and has absolute decision-making authority and financial responsibility. Thus, there was never any doubt during the Polaris program that the customer was the U.S. Navy, which made all decisions and footed the bill. The same applied for the Minuteman intercontinental ballistic missile, except that the customer was the U.S. Air Force. The objective also was clearly defined: produce a missile system that would meet certain specifications as to performance within certain fixed monetary resources by a certain time.

How would the systems analyst approach the task of eliminating pollution in Lake Superior? Using the systems methodology to isolate the problems and establish objectives presents no insurmountable difficulties. However, the analyst ultimately is faced with the question, "Who owns the lake?" Perhaps it could be stated a little differently from the management entrepreneur's viewpoint: "Who is going to pay for the systems approach to doing away with the pollution in Lake Superior?"

Here, there can be no clearly established customer-contractor relationship. The lake plainly is a complex thing when one begins to consider the ramifications of international, federal, and state laws regarding it. While United States or Canadian laws may obtain over navigation, for example, the states and provinces along its shore claim varying portions of its bottom or banks as their territory. Where does one go for the money and the authority to come up with a program that will result in a pollution-free lake?

Identification of the customer seems to be the greatest stumbling block to companies with an established capability in systems management. Once one leaves the realm of hardware and enters fields in which a multiplicity of regulatory bodies have overlapping and sometimes conflicting jurisdictions, such very abstract elements as politics are often the major considerations rather than time and money. Undoubtedly many more environmental and sociological problems could be solved by the systems approach if it were possible to clearly identify the responsible entities.

However, the structure of any government grows not logically and systematically but pragmatically in response to political expediency. There is no periodic review and reorganization of it on a functional basis. Thus there can be no single clearly established authority responsible for solving problems on a large scale. While one agency is paying farmers to leave land uncultivated, another is spending money to open new lands in arid areas for farming. While one agency seeks to eliminate segregation, another reinforces it by clearing out wooden ghettos and then building concrete ghettos in the same place, perpetuating the racial or economic distinctions that have prevailed for years.

The one governmental agency instituted so far during the space age that seems to have an organizational and financial responsibility to deal with a particular problem is the Appalachian Regional Commission. This body, which includes state and federal members, possesses the authority and funds to act in ways it sees fit to solve the problem of Appalachia. Obviously the commission represents a potential customer for the services of a systems-oriented management. However, the market is not so inviting as one would expect. For one thing, the commission has no guaranteed existence. It is funded, like other governmental agencies, on the whims of Congress. It will see lean years and fat years, in consonance with the degree to which Congress finds it politically expedient to support the objectives of the commission.

With the exception of the Appalachian Regional Commission and the Demonstration Cities and Metropolitan Development Act of 1966, Congress seems slow to adopt the systems approach or advocate it for agencies it creates and finances. For each bill passed that promotes the systems approach, many a deserving bill languishes and dies in committee because both the House and the Senate cling to the century-old philosophy of pork barrel politics as an answer for sociological problems. In 1967 and again in 1968, for example, Congressman F. Bradford Morse (R.-Mass.) introduced a bill calling for the establishment of a National Commission on Public Management, which expired in the House Government Operations Committee. The same committee failed to act on a similar bill (H.R. 312), introduced by Representative George E. Brown, Jr. (D–Calif.), in 1967, which would have authorized $125 million for grants to "the States and arrangements with institutions for the purpose of utilizing systems analysis and systems engineering approaches to national and local problems which have a substantial relevance to problems in other States."

The only valid test of how well the systems approach can work for essentially sociological problems may be simply to try it on a scale large enough to see tangible results. John H. Rubel, a systems engineer with Litton Industries, Inc., suggests that a completely new city be built using the systems approach from start to finish. It might cost billions of dollars and take as long as 10 to 15 years to realize, but the lessons learned could well set the pattern in urban renewal for centuries to come.

A project of such scope may furnish the only conclusive evaluation. As John Lessing points out in the January 1968 issue of *Fortune:*

> . . . if billions can be committed to a supersonic aircraft program, designed to reduce travel times for that fraction of the population that can afford it, why cannot billions be committed to a demonstration for the benefit of a whole people that city life can be made more worth living? Such a demonstration project would spin off innumerable immediate side benefits, ideas, and transferrable developments—not the least of which would be the promise that great city problems can indeed be solved.

4
OBSERVING THE EARTH FROM ORBIT

FOR PRACTICAL REASONS, OR SIMPLY TO satisfy an unexplained sense of curiosity, man for millenia has sought height by climbing mountains and cliffs so that he might gaze down at the Earth's broad lowlands and deep valleys. As he gradually progressed toward civilization, he became dissatisfied with the limitations imposed upon him by the regional topography and began to dream of soaring high into the skies like the birds. In legend, and then in fact, he attempted to fly. But, alas, without success: the human body wouldn't adapt itself to artificial wings, and the ideas of Daedalus and Icarus; of Bladud, the ninth-century "Flying King" of Britain; and of countless others were doomed to failure.

Where muscles failed, the brain succeeded. After centuries of dreaming, pondering, and experimenting, man finally took to the air in balloons. The Montgolfier brothers, Joseph Michel and Jacques Etienne, made the first public demonstrations of a hot-air balloon in June 1783 at Annonay, France. This was followed, on 19 September, by a flight to 1,500 feet by another balloon, this one containing a sheep, a cock, and a duck. The brothers were cautious: animals first, then man. Less than a month later, on 15 October, a 60,000-cubic-

First aerial photograph of the Earth, at left, obtained in 1906 from a camera installed in a rocket by the German rocket pioneer Alfred Maul. Below, an infrared photograph of the Earth, taken in 1957 by Major David G. Simons from approximately 100,000 feet in the Manhigh 2 free-flight balloon. (DEUTSCHEN MUSEUMS; ARFOR PICTURE ARCHIVES)

foot-capacity Montgolfier balloon supporting a "wicker gallery," and tethered by a rope to the ground, carried J. F. Pilâtre de Rozier to a height of 84 feet—the world's first captive ascent with a human aboard. And, on 21 November, the pioneering aeronaut and his companion, the Marquis d'Arlandes, made the world's first free flight into the air. They reached a height of 300 feet and came down some 5½ miles from their takeoff point, the gardens of the Château de la Muette.

Scientists quickly recognized that the new balloons would be ideal carriers of instruments to measure the nature of the air and of the winds beyond the normal reach of man. The first ascent of scientific instruments took place on 30 November 1784, when Dr. John Jeffries (an American physician and balloonist) and Jean Pierre François Blanchard (a French inventor and aeronaut) carried aloft six vials, together with a thermometer, a barometer, an electrometer, a hygrometer, a mariner's compass, and what was described by Jeffries as "a very good telescope." This early international aerospace experiment—the vials had been loaned to them by the English physicist-chemist Henry Cavendish to collect upper-air samples—produced the first scientific analysis of the composition of the upper air, and was a harbinger of things to come.

Balloons, and to an extent kites, were employed from the last quarter of the eighteenth century to loft scientific instruments, being joined during the twentieth century by the airplane and the rocket. As early as 1905 the potentiality of the last of these was recognized when Alfred Maul began obtaining photographs of the ground below by lofting a camera in a solid-propellant rocket. The first picture showing the curvature of the Earth is credited to Albert W. Stevens, U.S. Army Air Corps, who took it in 1935 from the Explorer 2 balloon at its peak altitude of 72,395 feet. David G. Simons, a major in the U.S. Air Force, spent more than 24 hours between 70,000 and 102,000 feet in the Manhigh 2 balloon in August 1957, and on 4 May 1961 Commander Malcolm D. Ross of the U.S. Navy ascended to 113,739 feet, a record for manned flight. Also during the 1960s, unmanned sounding and research balloons flew up to 150,000 feet (over 28 miles), rocket-powered airplanes to nearly 70 miles (NASA pilot Joseph A. Walker took an X-15 to 67 miles altitude on 22 August 1963), and military and atmospheric research rockets to hundreds of

Viking 2 photograph taken in May 1954 from an altitude of more than 150 miles, then a world's record. (U.S. NAVY)

miles above the Earth. Photographs from outside the atmosphere began to be made in the late 1940s by V-2 rockets, followed by Viking and other sounding vehicles, various military ballistic missiles, and, with the advent of the space age, from artificial, Earth-circling satellites.

ORBITAL OBSERVATION

The pioneers of astronautics recognized that orbiting vehicles would be ideal platforms from which to observe and monitor natural and man-made events as well as the oceans, the land, and the atmosphere. As their dreams blended into reality in the late 1950s and throughout the 1960s, it became clear that artificial satellites could be pressed into the service of mankind. From them, large stretches of the land and water surface across broad latitudes and longitudes could be observed simultaneously or nearly simultaneously—or according to a desired sequence under the same general conditions of solar or lunar lighting. Moreover, instruments in these craft could be pointed at areas difficult or impossible to reach by surface researchers: the Antarctic and Sahara wastes, the vast reaches of the oceans, the jungles of South America and Africa—even politically sealed-off nations such as China.

These synoptic (showing a wide-area distribution of conditions at the same time) observations of satellites can be duplicated only by dozens, hundreds, or even thousands of individual inputs from surface and airborne studies. And these same satellites can conduct long-duration measurements, over periods of months or years. Repeated surveillance over extended periods permits low-intensity, intermittent, and nonrecurring phenomena to be detected and studied. Moreover, movement and changes in intensity can be followed, such as the progression of snow lines and the deepening of seasonal coloring.

Observations from Unmanned Earth Satellites

Man was considered a vital component in the earliest studies of orbital stations, which were designed in part or wholely to observe the Earth's atmosphere, land, and oceans. However, these studies were made in the days before the advent of practical radio telemetry, a system that now permits measurements made in space to be transmitted remotely by radio to the ground. In the early 1950s, attention began to swing (temporarily, at least) from large, manned stations to small, unmanned artificial satellites, for the simple reason that telemetry had reached the stage where it had become feasible for use in such craft. In 1951, K. W. Gatland, A. M. Kunesch, and A. E. Dixon presented a study entitled "Minimum Satellite Vehicles" at the Second International Congress on Astronautics in London; and, at the fourth congress, S. Fred Singer offered his small Mouse concept, an acronym derived from Minimum Orbital Unmanned Satellite—Expendable (later, he changed the final word to "Earth"). Weighing but 100 pounds, it was to carry instrumentation to measure solar and cosmic radiations and magnetic fields in space. Singer concluded his proposal with the sage observation that in order to progress, one

> must be ready to justify a project even if the satellite is very small and minimal. Only in this way can we make use of the opportunities which the next few years may offer for such a project. If we can plan for this minimum satellite, the MOUSE, we may be launching it sooner than we now think possible.

Interest in small satellites continued, and in August 1955 Heyward E. Canney and Frederick I. Ordway presented a long survey, "The Uses of Artificial Satellite Vehicles," to the Sixth International Congress on Astronautics meeting in Copenhagen, Denmark. They emphasized that "scientific knowledge can be given a tremendous impetus following the establishment of even a small, unmanned satellite, and that lifetimes are such as to justify the existence of close orbit minimum vehicles." These and other proposals in the Soviet Union led to Russia's Sputnik 1, launched 4 October 1957, and America's Explorer 1 satellite, orbited in early 1958. Since the early Sputniks, Explorers, and Vanguards, spectacular progress with instrumented unmanned satellites has permitted many observational functions to be undertaken from space. Unmanned satellites have already added an immeasurable amount of knowledge about our planet and, together with manned vehicles, will continue to do so in the future.

In 1969, the U.S. National Academy of Sciences–National Research Council published a survey, *Useful Applications of Earth-Oriented Satellites*, backed up by 12 technical panel reports. Based on a study undertaken at the request of NASA, the survey gave priority to technological factors and the benefits that could be derived from the proposed on-board experiments to such fields as geology, geography, agriculture, forestry, meteorology, oceanography, navigation, and traffic control. The academy's Central Review Committee took great pains to (1) identify realistic space applications in the light of concrete needs within each discipline and of the benefits that might accrue; (2) assess the degree to which satellite-sensed phenomena or satellite-collected information could satisfy these needs; (3) determine the chief capabilities and limitations of platforms, vehicles, sensors, communication links, and data processors; (4) postulate feasible systems as a means of identifying tradeoffs (from such illustrative systems, some measure of understanding could be gained of the potential

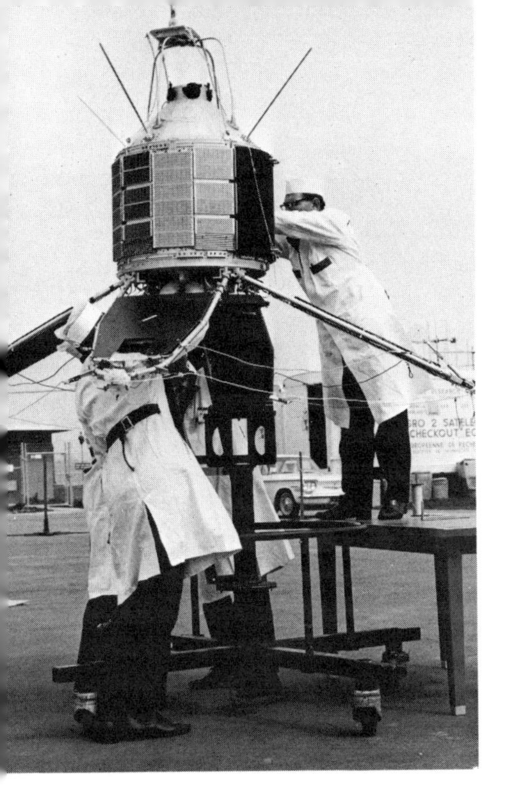

Typical scientific satellites. Clockwise from upper left: the British Ariel 3 (UK-3) undergoing checkout at NASA's Western Test Range in California; a typical configuration in the Soviet Kosmos series; a cluster of three United States Orbiting Vehicle satellites, OV 1-17, OV1-18, and OV1-19; testing of the American satellite Orbiting Geophysical Observatory 5 in a space simulation chamber. (NASA; ARFOR PICTURE ARCHIVES; U.S. AIR FORCE; NASA)

costs and timing that might be involved in satisfying each discipline's needs); and (5) identify problem areas that would require attention, for the individual application to be undertaken. The study showed clearly that there were so many common factors arising from orbital studies of the land, oceans, and atmosphere—such as orientation of the satellite, the resolution of the sensors, and orbital requirements—that careful design could produce "common-use systems" at costs significantly lower than if separate systems were developed for each of the many scientific fields investigated.

Observations from Manned Earth Satellites and Space Stations

Although the National Academy of Sciences study concentrated on unmanned Earth satellites to perform the many Earth resources missions and other activities, manned craft can also provide invaluable services to mankind. As the space age matures, it is likely that both will be employed for Earth resources work; only time will reveal what the scientific, technical, and economic tradeoffs are between the two. To date, manned satellites have complemented the work conducted by unmanned craft and have paved the way for scientific and practical operations to be undertaken in the orbital workshops, space laboratories, and space stations of the years ahead. The principal functions of man in orbit have been spelled out by members of the Science and Technology Advisory Committee and NASA in a 1969 report, *Uses of Manned Space Flight, 1975–1985*. In brief, they are (1) maintain, repair, and replace faulty subsystems; (2) update, modify, and replace subsystems; (3) deploy the initial structure; (4) align and calibrate instruments; and (5) operate instruments as a scientist-astronaut. When it is considered that nonfunctioning multimillion-dollar satellites might be made operable by an astronaut-technician—a single instance of the value of man in space—one appreciates that no matter how superb the satellite and its instrumentation may be, it can be very useful to have man around as a backup.

Probably the most immediate and obvious use of man in orbit is *service*—his ability to operate instrumentation in his own craft as well as retrieve, maintain, and repair unmanned satellites. The widespread presence of man in space will lead to increased modular design in

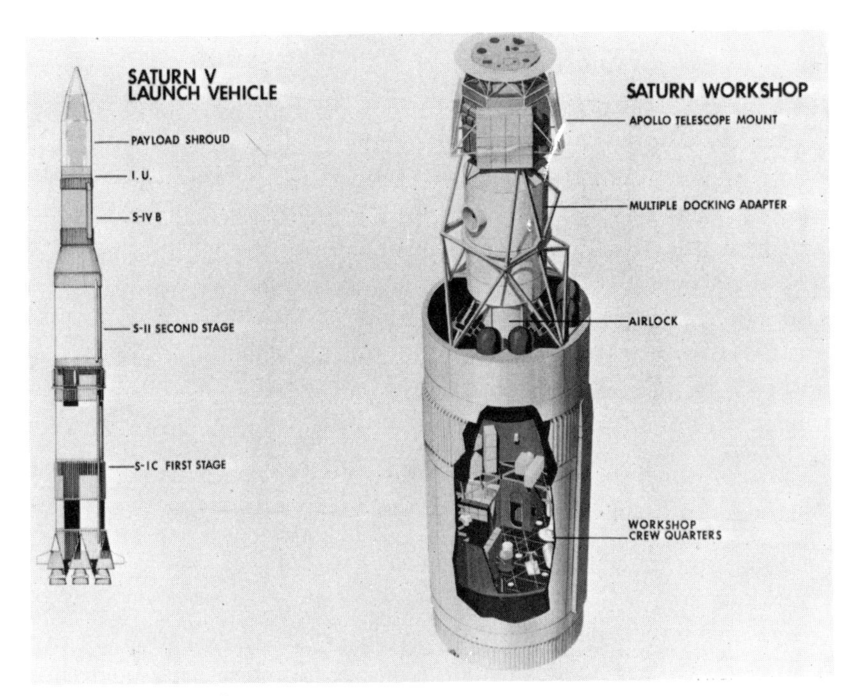

Large space stations planned for the future will enable men to perform complex experiments and observations over long periods of time. Above, the Skylab, shown attached to the Saturn 5 carrier vehicle, and with its major elements identified. Below, design for a 50-man space base; crew quarters and laboratory facilities at the ends of the telescopic booms would rotate on a hub around the stationary center section. (NASA)

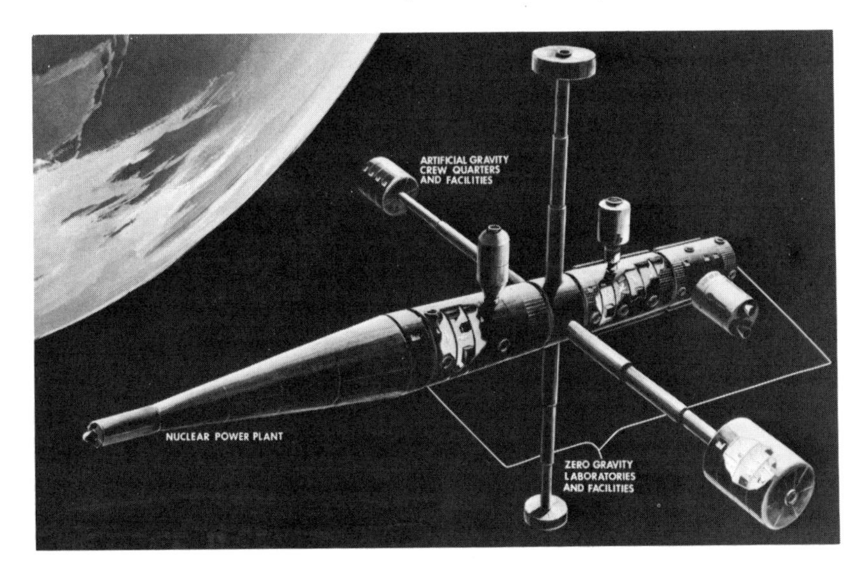

these satellites, so that malfunctioning subsystems can be readily replaced. Next in importance is the ability of man to serve as an *operator*, so that optimal use can be made of on-board instruments—for example, they may all be focused on a transient phenomenon such as a sudden solar disturbance or a major storm on Earth. Moreover, with a man on board immediate changes (for example, in instrument calibration) can be made in response to commands from scientists on the ground; and new instruments can be tested that are later to be flown in unmanned satellites. Man the operator may also control a number of unmanned satellites, and perhaps launch from his own craft other small satellites to carry out special tasks under his guidance.

As America's Skylab and other space stations or space station modules are orbited, and experience is gained, answers to many questions will be forthcoming. For example, is it possible to transmit effectively to the ground the total range of man's perceptions so that it can be analyzed rapidly and responsive data sent back up to the station? How important is immediate decision-making in terms of the observations being made? What are the effects of low- and high-orbit operations in terms of specific observations and feedback from the ground? How important might be the discovery of rare phenomena and unexpected situations? To what extent does two-way radio communication between the observer in orbit and the ground team influence the mission? What can man do that automatic sensors cannot? In his role as an on-board "sensor system," what sort of displays does he require in the space station?

As for manned spaceflight in general, in September 1969 the Space Task Group reported to President Richard M. Nixon (*The Post-Apollo Space Program: Directions for the Future*):

> . . . [the] manned space flight program permits vicarious participation by the man-in-the-street in exciting, challenging, and dangerous activity. Sustained high interest, judged in the light of current experience, however, is related to availability of new tasks and new mission activity—new challenges for man in space; . . . [his] presence . . . in addition to its effect upon public interest in space activity, can also contribute to mission success by enabling man to exercise his unique capabilities, and thereby enhance mission reliability, flexibility, ability to react to unpredicted conditions, and potential for exploration.

SATELLITES AND "THE USER" ON EARTH

Satellites, large or small, simple or complex, manned or unmanned, are considered not as individual vehicles housing sensing equipment but as the major element in an often complicated information-gathering system benefiting some "user." The system normally includes the launch site, the carrier vehicle, the satellite itself, tracking installations, ground receiving stations, and data links and processing centers. Occasionally the system becomes very complicated, as when a data-collection relay satellite is employed to locate, interrogate, and gather data from numerous information-generating platforms located in many widely separated, often remote, locations. Such platforms may be attached to balloons; airplanes; buoys at sea, on rivers, and on lakes; meteorological stations in the Arctic and Antarctic; or even migrating animals.

EARTH RESOURCES SATELLITES

Of ever-increasing importance to a rapidly growing community are special kinds of "user-oriented" or "applications" craft called Earth resources satellites. These orbiting vehicles can be so instrumented as to make a variety of studies, measurements, and observations of agricultural and forestry resources; hydrology and water resources; geology and mineral resources; geodesy, cartography, geography and man-made resources; oceanography and marine resources; and meteorology. The ultimate value of this new dividend of astronautics is impossible to foresee, but a study group established by the National Academy of Sciences has stated: "The potential economic benefits to our society from space systems are enormous . . . [they] may amount to billions of dollars per year to many diverse elements of our industry and commerce and use to the public. In some areas it is possible to predict these benefits with accuracy; in others we can estimate within broad but conservative limits." On the basis simply of what is known at the beginning of the 1970 decade, the U.S. Department of Agriculture estimates

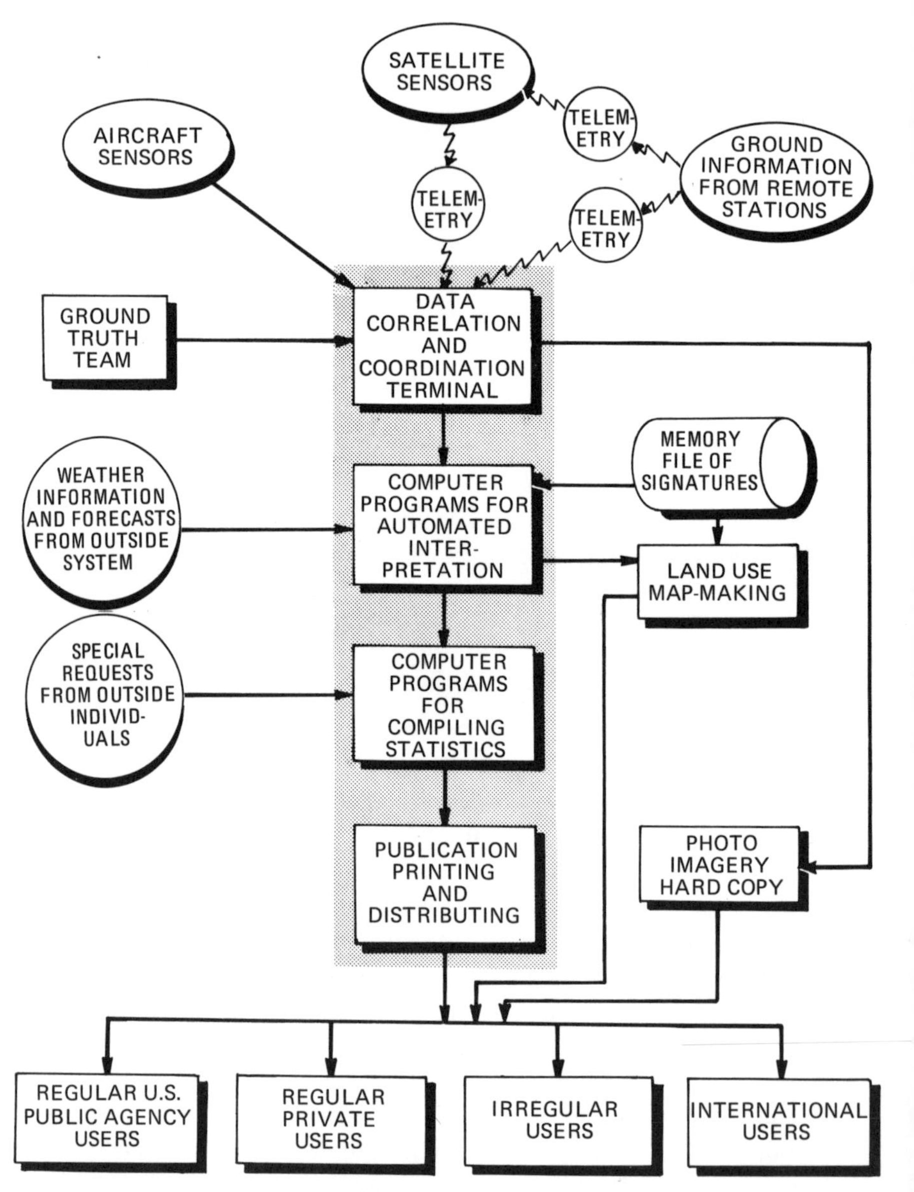

Earth resources information program for inventorying and evaluating productivity of the world's supply of food, fiber, and other natural resources, and to assess the interactions between man and his environment. (RICHARD P. MCKENNA)

annual American agricultural and forestry savings on the order of $3 billion a year to be derived from satellites fitted with imaging devices and other sensors. Such savings come from reduced crop losses due to satellite-derived weather and related data, the ability of satellites to detect and monitor the progress of crop diseases and forest fires, and the application of orbital information to assist in the assessment of optimum harvesting times, the probable size and quality of crops, and the like. These savings filter deep into the entire agribusiness enterprise.

In chapters 5, 6, and 7, the application of unmanned and manned satellites to the ocean, land, and atmospheric resources of the Earth is discussed. There is every reason to believe that orbital observation of our planet will be vital to the discovery of new resources, the management of dwindling reserves now known, and the detection and monitoring of air and water pollution.

The major unmanned satellite program dealing with these matters is that utilizing NASA's Earth Resources Technology Satellites (ERTS). A modification of the Nimbus spacecraft, the ERTS is instrumented to carry a specially constructed television system and a multispectral radiometer device that can acquire pertinent information on diverse species of crops, on plant diseases and plant vigor, on soil and rock types, and on their moisture contents. Coastal shoaling; ice, snow, and surface water distribution; and water and air pollution are also subject to study by this orbiting radiometric scanner, which can measure in both the visible and the infrared regions of the spectrum. (Radiometers are devices that measure either the density of total radiation or the radiation from a specific band in the spectrum, such as the infrared band.) ERTS spacecraft incorporate receiver-relay equipment as well, designed to collect information from remote data-gathering platforms on the land and ocean surfaces, and from balloons freely floating through the air.

Earth resources experiments conducted by ERTS and such other unmanned satellites as Nimbus, Tiros, Essa, Itos, Application Technology Satellites (ATS), and Geos, as well as the manned Mercury, Gemini, and Apollo series, rely to a great extent on photographic and other instruments that read, from their remote positions in orbit, "signatures" produced by the land, waters, and air down below.

SIGNATURES AND REMOTE SENSORS

Just as human beings can be identified by their fingerprints or voiceprints, the different features of the globe can be identified by their signatures in the electromagnetic spectrum. Objects of every sort and every size—from individual bare rock outcrops to whole deserts, from steaming jungles to vast Antarctic ice fields, and from small apple orchards to extensive wheat fields—absorb, reflect, and emit characteristic wavelengths from the ultraviolet through the microwave.

Signatures reveal the condition of objects as well as their nature. For example, the remote sensor on a satellite not only can distinguish ocean from land, it can sense the difference between rough water and smooth water or between warm water and cold water. This is because the radiation from any specific source varies according to its surface structure and temperature, among other factors. In agriculture, to take another example, each plant has a distinct "slot" in the spectrum. When disease strikes trees, they lose chlorophyll, and this change causes a shift in the signature of the unhealthy trees. In an infrared photograph of the forest, the diseased trees will stand out darker than the healthy ones. In the same way, observations in the near-infrared can lead to the detection of wheat fields that are plagued with wheat rust or suffering from inadequate fertilization or irrigation.

By measuring electromagnetic signatures, Earth resources satellites collect great volumes of information in a fraction of the time that would be required by land-based observers. This means, of course, that decisions of all sorts—from the plotting of an ocean liner's course to the cutting out of diseased trees—can be made with hitherto unimaginable speed and accuracy. Problems can be spotted and attacked before they have grown to such proportions that they are no longer susceptible to easy solutions.

Earth resources satellites carry multiple sensing devices, including photographic systems, radar, and radiometers. Agricultural and forestry users concentrate on the optical spectrum, especially the near-infrared region, while oceanographers employ somewhat longer wavelengths associated with thermal radiations from water bodies. At

TABLE I

TYPICAL APPLICATIONS OF REMOTE SENSORS

SENSING TECHNIQUE	AGRICULTURE AND FORESTRY	HYDROLOGY	GEOGRAPHY AND GEODESY	GEOLOGY	OCEANOGRAPHY
Photography, Visual	Crop and soil identification Identification of plant vigor and disease	Identification of drainage patterns	Urban-rural land use, transportation routes and facilities	Identification of surface structures	Identification of sea state, beach erosion, off-shore depth, turbidity along coasts
Photography, Multispectral	Crop and soil identification Identification of plant vigor and disease	Moisture content of soils	Terrain and vegetation characteristics	Identification of surface features	Sea color as correlated with productivity
Infrared Imagery and Spectroscopy	Terrain composition Plant vigor and disease conditions	Detection of areas cooled by evaporation	Surface energy budgets Near-shore currents and land use	Mapping of thermal anomalies Mineral identification	Mapping of ocean currents Investigations of sea ice
Radar Imagery	Soil characteristics	Moisture content of soils Identification of runoff slopes	Mapping of land ice Cartography Geodetic mapping	Surface roughness Tectonic mapping	Sea state, ice flow, and ice penetration Tsunami warning

TABLE I (continued)

TYPICAL APPLICATIONS OF REMOTE SENSORS

SENSING TECHNIQUE	AGRICULTURE AND FORESTRY	HYDROLOGY	GEOGRAPHY AND GEODESY	GEOLOGY	OCEANOGRAPHY
Radiofrequency Reflectivity	Soil characteristics	Moisture content of soils	Land ice mapping and thickness of land ice	Measurement of subsurface layering	Sea ice thickness and mapping
					Sea state
			Penetration of vegetation cover	Mineral identification	
Passive Microwave Radiometry and Imagery	Brightness temperature	Snow and ice surveys	Snow and ice measurements	Dielectric constant	
	Mapping of terrain			Measurement of subsurface layering	
Absorption Spectroscopy				Detection of mineral deposits	Detection of concentrations of surface marine flora
				Trace metals and oil fields	

Adapted from NASA Headquarters Manned Space Sciences and Applications, Washington, D.C.; and NASA Manned Spacecraft Center, Houston, Texas.

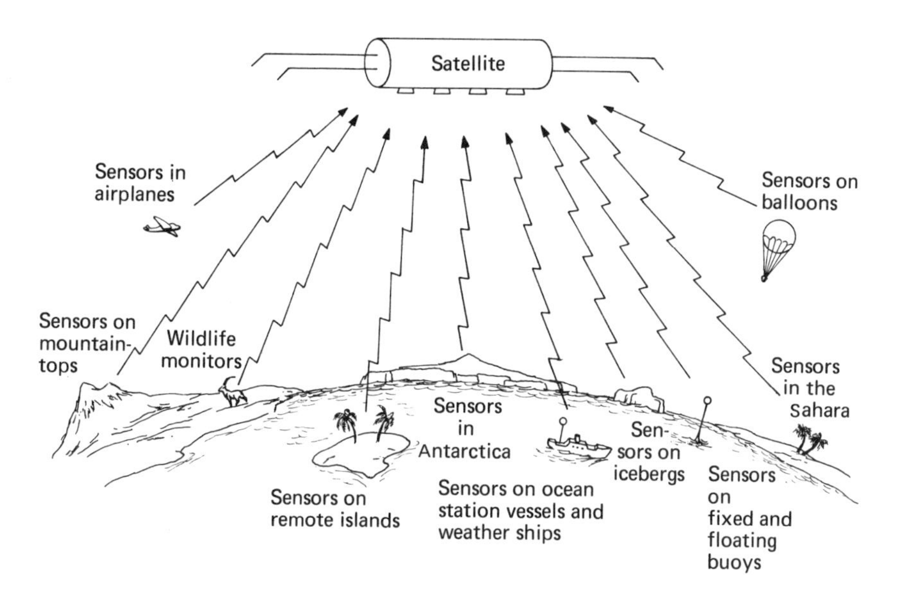

The Earth resources satellite is a highly versatile device, capable of collecting tremendous amounts of data across a wide range of the electromagnetic spectrum. Sensors can be attached to almost anything, from a buoy for monitoring the meanderings of the Gulf Stream, to an elk for studying its migratory route. (RICHARD P. MCKENNA)

still longer wavelengths, in the microwave spectrum, passive radiometers quantify such phenomena as the temperature of the soil, while active microwave radar systems can perform ocean and land topographic measurements, their returned signals yielding information on the roughness of the surface being probed.

The most important recording devices used for Earth resources surveys are photographic systems. For mapping, the frame camera is most suitable, but for high-velocity, low-altitude use adequate compensation for satellite motion must be built in. Slit cameras (whose film passes a narrow slit in the image plane) automatically compensate for the motion produced in the image, while the panoramic camera employs rotating mirrors, lenses, and/or prisms to permit "horizon-to-horizon" photography. Panchromatic is the most widely used photography because of its ready availability, low cost, and the broad experience in its interpretation. The technique of multispectral photography relies on the correlation of each element of a scene with the characteristic reflectance of the material for which the sensor is searching. A background of spectral signature studies of land

and water features is necessary for the multispectral technique to be successfully employed.

Infrared techniques are particularly useful in making measurements of temperature changes of both land and water surfaces, taking advantage of the fact that all material things at temperatures above absolute zero emit infrared radiation within the electromagnetic spectrum at wavelengths between 0.7 micron and 1,000 microns (1 mm). The detectors employed to measure this radiation are passive; that is, they do not emit a pulse (like radar) toward the object being sensed. The infrared instrument scans the surface below and the results are displayed on a cathode ray tube, whose face is photographed by the recording camera to produce the permanent record.

Unlike infrared radiometers, passive microwave sensors can be used in all weather conditions; they can "see through" clouds, are of relatively small size, and have low power requirements. Since they measure brightness temperatures, they are most useful in mapping ocean and snow surfaces.

Radar, also operating within the microwave part of the spectrum, is an active sensing technique using electromagnetic energy in the wavelength band between 1 mm and several meters (its most useful range being 0.5 to 15 cm). In an active, as opposed to passive, system, the response of the returned signal is largely a function of the reflectance of the target. Radar imaging has been used for several decades in aircraft, first with plan position indicator systems and later in side-looking radar systems. The latter transmit to the area to be illuminated below from antennas located on each side of the carrying craft. The antennas also act as receivers, detecting the return reflections of the energy originally transmitted. The returns are displayed on a cathode ray tube, which is photographed to provide the permanent record. The wavelength chosen depends on what is to be sensed; shorter wavelengths are scattered, attenuated, and absorbed by atmospheric water vapor, while longer wavelengths more easily penetrate the atmosphere and impinge on the surface. Because water surfaces are good reflectors of microwaves, a uniform, black-tone image results, useful in identifying drainage systems. Radar's ability to penetrate foliage makes it useful in enhancing the topographic profile beneath.

Infrared measurements of heat reveal far more than the human eye can see. A drop in the temperature of corn, for example, may indicate the presence of blight long before it is apparent to the farmer. Here, an engineer plays back the tape from the test of a multispectral scanner designed to survey the Earth from an airplane or a spacecraft. (THE BENDIX CORP.)

As a means of avoiding undesirable variations in light intensity, shadows, and edge effects, Sun-synchronous orbits are used for Earth resources satellites. Such satellites enjoy the same Sun angle throughout the year as they orbit the Earth; that is, the orbital rate of precession matches the angular rate of the planet's movement around the Sun.

The altitude at which a particular Earth resources satellite orbits is always a balance between such factors as the desire for the closest view possible, wide geographic coverage, the optimum time required to make a given observation, and the demands for rapid collection and relay of data, precise attitude control, and accurate satellite station-keeping—that is, the ability to keep the craft in a given orbit at a given orientation. Assuming that an Earth resources survey satellite is placed in a 500-mile-high, Sun-synchronous orbit, it will be able to observe and photograph any individual target area some 20 times each year, since its orbital period around the Earth will be approximately 100 minutes, during which time a target on the Earth beneath will rotate about 1,500 miles.

EXPERIENCE IN EARTH RESOURCES SENSING

In the Mercury, Gemini, and Apollo manned spacecraft programs, systematic terrain photography was undertaken with a remarkable degree of success. Some 60 useful Earth resources photos were obtained during the Mercury program, and 1,400 in the Gemini program. In the fifth and sixth Earth-orbital unmanned Apollo missions, cameras were bolted to the window frame of the command module, while in the manned Apollo 7 a camera was mounted outside, facilitating vertical and near-vertical photography. In early March 1969, the NASA multispectral photography experiment SO65 was flown for the first time, aboard the Apollo 9 satellite. Four 500-EL Hasselblads installed on a common mount and synchronized for simultaneous exposure were employed.

The astronauts were highly enthusiastic about the photography experiment, as a fantastic panorama unfolded before their eyes. Colonel James A. McDivitt reported

> . . . a tremendous picture of a weather system—a cyclonic thing that was actually seen from the weather satellites [Essa unmanned meteorological satellites] in only a matter of three to four hours after this [moment of Apollo 9 photography]. By correlating this photograph with what was seen on the transmission from the weather satellite, we were actually able to show, when you get this glob of black and white from the weather satellite, what it really looks like from orbit, and what the weather man can do with that. When he sees something like this, he's pretty sure what it is. But sometimes the weather satellites aren't quite so clear. So there's a good correlation between these two [the Apollo 9 photograph and one produced by the unmanned system].

Most of the photographs resulting from manned satellite cameras provided excellent clarity and detail and have been put to good meteorological, oceanographic, and land resources use. For unmanned satellites, high-resolution television cameras—of the type developed for the Application Technology Satellite (ATS), Tiros, Essa, and Nimbus Earth-orbiting series as well as for lunar and planetary spacecraft —have supplied additional, and long-term, coverage of the Earth below.

Beginning in late 1964, NASA introduced into service a Convair

240A aircraft as a test bed for remote sensors to be used in Earth resources satellite programs. Several years later, a Lockheed P-3V joined the program, followed by a Lockheed C-130B in 1969. Among the instruments flown are mapping cameras, multiband cameras, infrared scanners, ultraviolet scanners, microwave radiometers, microwave imagers, and radar scatterometers. All data resulting from the flights are processed by the Manned Spacecraft Center in Houston, and supplied to the principal "user" agencies: the Department of Agriculture, the Department of the Interior, and the Naval Oceanographic Office. Excellent experience has been gained from these flights not only in how to handle the data coming in and how best to make use of them, but in determining what measurements can, in the interim, best be made from aircraft pending full sensor development for satellites. An interesting by-product of the remote sensing aircraft program was the use, in July 1970, of a Lockheed Electra that was dispatched by NASA to northern Peru to assess the damage caused by the great earthquake and to help the Peruvian government plan for reconstruction. It also provided invaluable experience in applying the science of remote sensing to natural catastrophes.

Although primarily meteorological satellites, the Tiros and Nimbus craft have also yielded useful oceanographic and land resources data. In the first 10 years of the meteorological satellite program, 10 Tiros satellites (whose name is derived from Television InfraRed Observation Satellite), 9 Essa satellites, 3 Nimbus satellites, and 1 Itos (Improved Tiros Operation Satellite) were successfully orbited. (Tiros 1 started the program on 1 April 1960.) Also placed into operation during this decade were two ATS missions with meteorological objectives; they were put into a stationary geosynchronous (24-hour-period) orbit 22,300 miles high. In all, more than a million pictures of the Earth's cloud cover, surface, and other data were produced.

The Nimbus program has been especially broad. From the point of view of geology, photos have allowed the U.S. Geological Survey to correct errors in Antarctic maps, locate new geographic features, and provide a deeper understanding of such well-known features as the Paris Basin in central France. By photographing the limits of snow cover, qualitative estimates of snow cover between ground stations were made, useful in predicting groundwater runoff and making flood estimates. Much ice pack reconnaissance was accomplished,

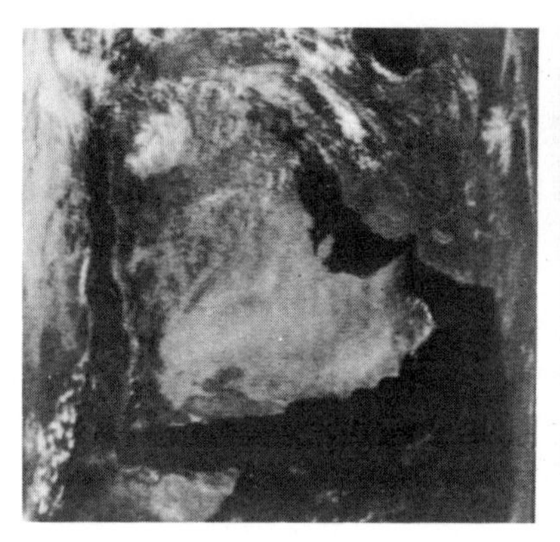
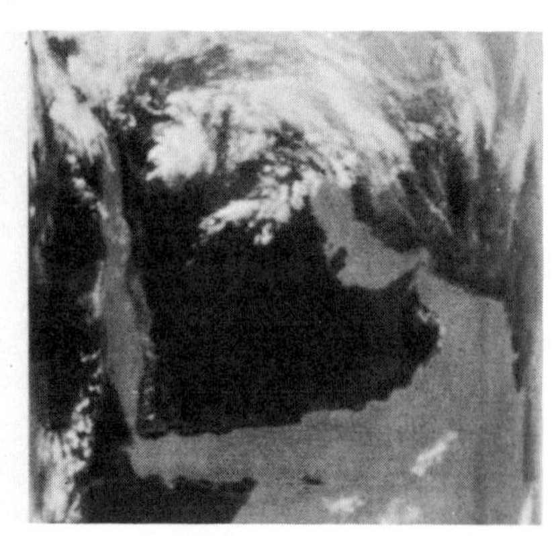

Itos 1, launched in January 1970, has provided pictures of cloud cover at night as well as during the day. The photograph of the Arabian peninsula at left was made by daylight; the one at right, in infrared during total darkness. (NASA)

particularly in remote areas where little knowledge was available. For the first time, an iceberg (some 70 by 20 miles in size) was "photographed" by TV and infrared radiometers over the polar regions; and, during the course of many months, ocean circulation and sea-ice drift measurements were made regularly. Nimbus satellites were also able to determine, by infrared photography, various boundary points along the Gulf Stream–ocean interface.

Nimbus 3 carried an infrared interferometer spectrometer whose objective was not only to provide information on the vertical structure of the atmosphere but to measure thermal emissions from the surface. The spectra derived from the instrument were useful in furthering research on ocean surface temperatures, observing from orbit residual ray phenomena in desert areas where minerals may be exposed, and working out ways to detect and monitor atmospheric pollutants. The satellite also carried in Interrogation, Recording, and Location System (IRLS), designed to identify, locate, interrogate, and store data from remote platforms in the air (aircraft- and balloon-borne), on land (for instance, weather stations, and an elk!), and at sea (buoys and a floating ice island). The Bureau of Commercial Fisheries, for example, placed a drifting buoy in the northern Pacific Ocean south of Alaska to measure water surface temperatures, water temperature and pressure at a depth of 165 feet, and the salinity of seawater 3 feet below the surface. Nimbus 3 also proved capable of tracking ships at sea with a high degree of accuracy.

NAVIGATION SATELLITES

Ships at sea can do the opposite, and track satellites in the skies. For example: in the early morning hours of 24 November 1969, the aircraft carrier USS *Hornet* was cruising through the Pacific Ocean, some 400 miles southeast of American Samoa. Her mission was to be on station when the Apollo 12 spacecraft, approaching from the Moon, splashed down, and then to handle the recovery. The Yankee Clipper, Apollo 12's returning command module, was scheduled to land 244 hours, 35 minutes, and 23 seconds after takeoff from Cape Kennedy on 14 November. It arrived 1 minute and 2 seconds late at 15 degrees 47 minutes 6 seconds south latitude and 165 degrees 5 minutes west longitude–with the *Hornet* on station as scheduled.

That the *Hornet* was in the right place at the right time was no mean feat of navigation; and it could not have been accomplished without the aid of a navigation satellite some 610 miles above the ocean in a near-polar orbit. As the *Hornet*'s skipper, Captain C. J. Sieberlich, said on the occasion:

> . . . the value of the navigation satellite [Navsat] during the Apollo 12 recovery operations cannot be overstated. . . . *Hornet* was required to arrive at the Apollo 12 splashdown point four hours prior to splashdown. . . . She had to be steaming through a point five nautical miles north of the splashdown point headed south at the exact time the main parachutes of the Apollo command module opened. During the five hours prior to the splashdown, four NAVSAT fixes were obtained. . . . The results achieved point out the accuracy of the system. *Hornet* was able to reach a point only six thousand four hundred yards from the command module, Yankee Clipper, by the time it descended to the surface.

Such precision in navigation would have been impossible using other means. Loran (LOng Range Aid to Navigation) and bottom-contour navigation are not available in the South Pacific. In addition, the last star fix was taken at a position some 180 miles from the recovery point. As the *Hornet* steamed on, the day dawned cloudy and overcast, preventing subsequent celestial observations. Dead reckoning techniques were ruled out by high winds and heavy seas.

The Navsat program went into operation on 30 April 1960, when the U.S. Navy orbited Transit 1B, the world's first navigation satellite. Over the next few years, other navigation satellites were launched by the United States as well as by the Soviet Union. By 1965 both countries had operational systems, used primarily for the benefit of submarines and a few of their larger surface ships.

The Navsat system consists of a number of satellites in circular, near-polar orbits at some 600 miles altitude. Once each 12 hours, a new orbit is calculated by Earth tracking stations for each of the satellites, and the resulting data are stored in the satellite's memory unit. The satellite then transmits the orbital data at 2-minute intervals. The navigator on the ship determines when the satellite is directly overhead by analyzing its radio signal, which changes frequency as the satellite approaches, then recedes from, the vessel (the Doppler effect). This information, together with the orbital data, the time the signal was received, and the speed of the ship, then is fed into an on-board computer, which rapidly calculates the ship's position. The system is accurate to within a fraction of a nautical mile.

In July 1967, the U.S. Navy made the Navsat system available to civilian shipping, but there were few takers. The cost of the shipborne computer and associated communications equipment was simply too great for all but specialized craft devoted to oceanographic research and oil exploration. In addition, the sophisticated equipment demands highly skilled and highly paid personnel to operate and maintain it.

However, one civilian ship, the *Queen Elizabeth II*, which completed her maiden voyage by sailing into New York harbor on 7 May 1969, was designed with the Navsat system in mind. For a passenger ship that cost $72 million, the $65 thousand for Navsat computer and on-board receiving equipment did not loom very large. Cruising at 28 knots, the 58,000-ton liner was kept on course during her Atlantic crossing with a positional error of only a few hundred feet. Antennas on her foremast received data from the Navsat system and transmitted them to a small computer in the ship's chart room. In only 20 seconds the space-age navigator had his position neatly typed out.

Like so many other dividends from space research, the navigation satellite remains largely unexploited because of the costs involved in adapting it for nonmilitary uses. But the potential market for it in the shipping industry is great. Undoubtedly, continued improvements in the state-of-the-art will make the system more economical, and the future will see it more widely utilized in civilian shipping.

AIR TRAFFIC CONTROL SATELLITES

Functionally akin to the navigation satellite is the air traffic control satellite. Indeed the two could be combined quite easily. Unlike the former, the latter has undergone little development despite the obvious need.

As the second decade of the space age began in the United States, air traffic controllers of the nation were staging "sick-ins" to call attention to their plight and the lack of foresight on the part of the federal government; jumbo jets were rumbling off the nation's runways; and the supersonic transport was moving into production in the Soviet Union, France, and England—while the United States continued to fall further behind in the aeronautical technology of the future. Meanwhile, the concepts of the V/Stol (Vertical and Short Take-Off and Landing) aircraft were proving economical for the short-haul and commuter segment of the air passenger traffic, particularly in Europe, and helicopter "taxis" from metropolitan airports to downtown landing pads were further crowding the sky. However, instead of developing a coherent air traffic control system, the United States continued to rely on a patchwork quilt of radio beacons and VOR (Very high-frequency OmniRange) stations marking routes between airports.

The reasons for adhering to the existing system, even though it is not adequate to handle the increasing demands being placed upon it, are many and complex. The airlines, for example, are reluctant to subscribe to any new system that would involve complete replacement of the costly electronic equipment now used in aircraft and ground stations. Moreover, there have been disputes within the industry over the exact role of the proposed satellite system. The International Civil Aeronautics Organization has said that an aeronautical traffic control satellite is needed primarily for communications and surveillance, not for navigation—a position generally endorsed by airline pilots and many air traffic controllers. The National Academy of Engineering, however, has recommended that NASA be given responsibility for the development of new technologies in air traffic control, with particular emphasis on satellites for more accurate navigation. The President's Scientific Advisory Committee panel on air traffic control also has urged that satellites be used to replace radar and that more automated procedures be adopted in order to reduce the need for human traffic controllers. On the technical side, decisions still have to be made whether to use VHF (Very High Frequency) channels of communication, which are rapidly becoming depleted, or UHF (Ultra-High Frequency) channels, which have been allocated for future needs in the field of aeronautical communications.

Matters were resolved early in 1971 when the President's Office of Telecommunications Policy recommended the early deployment of aeronautical satellites operating in the UHF range and providing both air-ground communications and air surveillance service. Significantly, too, the recommendation suggested the use of "commercial telecommunications facilities and services to the maximum extent feasible in both pre-operational and operational aerosat [aeronautical satellite] systems." Thus the aerosat will clearly be owned and operated by private enterprise rather than by the federal government. The same recommendation also cleared up the problem of which federal agency would be in charge of the aerosat program in its developmental phase. It clearly spelled out that the Department of Transportation, that is, the Federal Aviation Agency, would be the "lead management agency." NASA's role would be to "conduct independent research and development on technologies which have broad application and, under the management and budget of the Department of Transportation . . . provide other technical support unique to transportation applications."

Plans call for the orbiting of a preoperational satellite over the Pacific in 1973 and a similar one over the Atlantic in 1975. A completely operational system with the participation of other nations is seen for the 1980s. The future, then, of the aerosat is in the hands of the entrepreneur rather than a federal agency. Whether the ambitious time frame can be met will depend upon the degree to which private capital is available for such a project and the degree to which it is supported by funds from a federal agency that has, admittedly, more pressing needs than satellites.

The airlines themselves began making tests concerning aircraft and communications satellites in 1965; in 1967, seven airlines conducted 125 hours of such experiments using NASA's ATS system. But experiments in navigation received a setback in August 1969, when ATS-5 began tumbling in orbit and could not be stabilized. The U.S. Department of Defense also sponsored studies, beginning in 1968, of an advanced air navigation satellite (Project 621B) by Hughes Aircraft Company and TRW, Inc. These grew into a proposed defense navigation satellite system, intended primarily for tactical weapons delivery for all military services. As 1971 opened, however, funds for hardware were severely limited.

Thus the air traffic control satellite is another example of a valuable space dividend that has not been realized. Though the reasons for this are essentially nontechnical, even if sufficient funds were made available today it probably would require ten years to design, produce, and test such a complex system. However, the time lapse could give designers enough margin to provide a satellite that would do a multitude of jobs in bringing order and greater safety to the crowded skies. Such a satellite could provide operational communications (airline message traffic), navigation, traffic control, collision avoidance, passenger telephone service, weather advisories, and search and rescue data. For the high-flying, supersonic transport it could, in addition, provide warnings of solar-flare events in time to permit the pilot to dive deeper into the atmosphere, to protect passengers and crew against the resulting radiation.

Such a satellite could also be used by ships, eliminating the need for two navigational systems. Space experts of TRW, Inc., feel that the air traffic control and navigation satellite of the future could well establish the location of aircraft or ships within a positional accuracy of 60 feet, and their velocity to within 0.2 fps. In addition, aircraft altitude could be computed to an accuracy of 130 feet absolute, and rate of climb to within 0.4 fps. The satellite would provide these services continuously, or nearly so, regardless of the weather. It would weigh only 350 pounds and remain in its Earth-synchronous orbit for at least 5 years.

With such a satellite technically feasible at the beginning of the 1970s, the nation, and the world, possesses a space dividend that could be realized by 1980—or sooner.

GEODETIC SATELLITES

Satellites are especially applicable to geodesy, the science that deals with the shape of the Earth and the structure of its gravitational field. Such satellites have been developed to improve land survey networks and reference points, relating these points ever more accurately to each other and to the center of the Earth. These satellites can also help to establish the true shape of the Earth, as well as variations in thickness and density of its crust that cause anomalies in its gravitational field.

Geodetic satellites are either passive or active—the former being merely reflectors of sunlight or radar signals, while the latter have some form of transmitting equipment on board. Typical of the passive satellite is Pageos 1, an aluminized plastic balloon only 0.0005 inch thick and 100 feet in diameter. It was launched on 23 June 1966. By photographing this satellite against the star background, positions on the Earth were charted with extreme accuracy. For example, scientists of the Japanese Hydrographic Office used this method to calculate the position of the island Tiro Shima, some 350 miles south of Tokyo. Photographs of Pageos 1 (and Echos 1 and 2, as well) showed them that the island was in reality 1,640 yards east of where it was shown on naval charts. Russian cartographers and geographers also used photographs of the orbiting Echos 1 and 2, in 1961, to place the city of Kharkov on their maps with an estimated error of only 50 yards.

Active geodetic satellites—much smaller and more sophisticated—are used today rather than the passive ones. The first of the active satellites was Anna 1 (an acronym indicating its sponsors: Army,

Navy, NASA, Air Force), launched on 31 October 1962. Anna 1 was orbited from Cape Kennedy (then Cape Canaveral) and had on board two xenon tubes that produced a series of 5 flashes 5.6 seconds apart some 20 times a day. The flashes showed up as a dotted line on photographs of the satellite's path across the night sky. In addition, Anna 1 transmitted radio signals to permit position-finding by Doppler techniques. The use of the flashing light and Doppler shift was a great improvement over reliance on reflected sunlight, since they provided more flexibility for satellite trackers over a wider portion of the Earth's surface.

Current active satellites include the Secor series—compact, rectangular boxes 9 by 11 by 13 inches that weigh only 45 pounds. They are powered by solar cells. Secor is so small and lightweight that it can go into orbit very economically by "hitchhiking" a ride with a larger satellite. Sponsored by the U.S. Army, the Secor (SEquential COllation of Range) series has been used extensively in a program of mapping islands in the Pacific Ocean with extreme accuracy. To communicate with the Secor satellites the Army developed mobile ground stations that can easily be transported among islands. Three are positioned at accurately known locations while one is put on the island (or other location) to be mapped. A 12-pound radio transponder in the Secor satellite receives signals from the stations at the three known positions and retransmits them. Equipment at the fourth station compares the phase and frequency of the return signal with that transmitted and computes the distance to the satellite with great accuracy. Similar comparisons of this type on subsequent orbits of the Secor make it possible to locate the fourth station very accurately with respect to the other three.

Thus the geodetic satellite provides a valuable tool for establishing a worldwide, three-dimensional reference system. In addition to generating data for highly precise maps of islands and continents on the Earth's surface, these satellites aid in the more accurate mapping of the bottoms of the seas and oceans.

5
THE OCEANS

MAN KNOWS LESS ABOUT THE MYSTERIOUS depths and floors of the oceans than about the Moon, although he lives on a world covered 71 per cent by broad expanses of water. Not without reason did Thomas Gray write of the "dark unfathom'd caves of ocean," and Thomas More of its "sunless retreats."

The oceans, and their contingent seas, sounds, bays, and inlets, influence man in many ways, from the climate he experiences, the oxygen he breathes, and the food he eats to the economic, social, political, and military conditions under which he lives. Four out of five of the planet's living creatures make their abodes in these huge, interconnected salt-water bodies that encircle the globe; indeed, some 200,000 species have already been identified in the oceans, and more are discovered yearly.

Impressive in volume as well as area, the oceans contain 330 million cubic miles of water, whose average depth is 12,450 feet (compared with a mean continental height of only 2,760 feet), spread over 140 million square miles of surface. The surface area of the Moon, by contrast, is only about 1/14 that of the *entire* Earth; and, already, virtually all of it has been photographed and mapped by Lunar Orbiters, Luna and Zond unmanned spacecraft, and Apollo manned spaceships.

In the past, the oceans have been studied largely by research ships slowly plying them. More recently, deep-sea submersible diving research vehicles and underwater experimental laboratories—such as Sealab III—have been developed, augmented by deep-sea data buoys (some of which are dubbed "Bumblebees") and ocean-bottom-mounted tidal instruments. In the study of the biological properties of the oceans, automatic recording devices are used to measure such things as volume, abundances of plankton particles, and concentrations of surface chlorophyll.

The sheer size of the oceans makes it extremely difficult and time-consuming to measure and to monitor even surface phenomena, much less gather adequate information at depth and from the floors. The costs of operating oceanographic research vessels run from $2,000 to $4,000 a day. Based on an average square-mile-per-day survey by a typical ship, some 14,000 ship-days would be required to survey the world's oceans at a cost of between $28 million and $56 million. In order to provide synoptic surveys that permit ocean conditions to be assessed over a wide region at a given time, the total initial cost would be $56 million, based on an estimated per-ship cost of $4 million. Today, only sporadic research efforts are carried out over selected portions of the oceans; for example, a program was undertaken by the Office of Naval Research, the Coast Guard, and the Scripps Institution of Oceanography to anchor numerous small buoys some 1,000 miles north of Hawaii to study the upper layers of the ocean and the interfacing atmosphere. While the program was successful, the process of visiting each buoy by surface ship and reading its instruments was painstakingly slow.

The practicability of using aircraft to survey selected areas of the ocean, particularly over coastal waters and along well-established intercontinental routes, has been proven. Radio-transmitting buoy platforms have been successfully interrogated by such aircraft, permitting definite improvements to be made in gathering data. Unfortunately airplanes probably never will be used to survey and to monitor long stretches of the ocean surface far from continental shorelines simply because of the enormous fuel requirements, delivery and replacement costs, and costs of operating and maintenance crews.

Still, the need for worldwide information on the oceans exists. As the space age matured, it became evident that there was one feasible

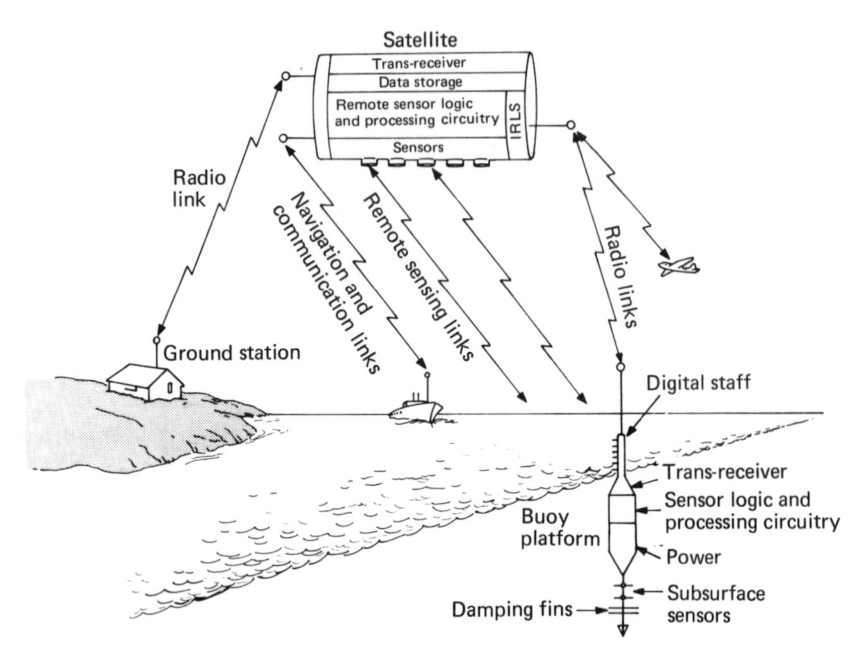

Besides gathering data, oceanographic satellites relay information through the Interrogation, Recording, and Location System. Data about sea state, for instance, might be obtained from sensors on an airplane or a buoy and transmitted to a station on land—or the information might be supplied directly to a ship below, enabling the captain to plot the best possible course. (RICHARD P. MCKENNA)

type of platform on which sensors could be placed that could solve the problem, a platform that in a relatively short period of time could view the oceans and seas all over the world. It was the artificial satellite.

The application of space science and technology to the study of the oceans began in earnest as the 1970 decade got under way. It appears certain that, as oceanographic satellites are introduced and experience is gained in their use, they will take over more and more of the research now accomplished by conventional surface and subsurface methods. Although satellite surveillance of the oceans is presently limited to the surface and immediate subsurface waters, spaceborne observations can make contributions to studies of the depths by pinpointing and interrogating moored or free-floating buoy platforms that temporarily store information gathered by surface and subsurface sensors. Of course, oceanographic-research and other vessels, including fishing trawlers, can also relay the data they acquire via oceanographic satellites. Data transmitted to the surface by deep-submersible craft can also be so relayed to ground stations for evaluation and dissemination.

OCEANOGRAPHIC SATELLITES

Like any system, the oceanographic satellite enjoys advantages and suffers limitations. On the plus side, satellites can monitor ever-changing ocean phenomena simultaneously or nearly so. If a stationary orbit more than 22,000 miles above the equator is selected, an entire ocean may be observed. For closer views, relatively low polar orbits (a few hundred miles) permit the same ocean to be surveyed at intervals of less than a day, acceptable for making many synoptic measurements.

A seemingly severe limitation of oceanographic satellites is the essentially two-dimensional picture they paint. In reality, however, such a picture contains much information of significance to man. It shows a surface where the atmosphere and the ocean meet, the interface of principal energy flow giving rise to storms, waves, and currents. Moreover, the daily tides occur in the outer layer of the ocean waters. Here photosynthesis takes place and here, too, most of the biological resources of the seas are found. Surface waters flow onto beaches and into harbors, support icebergs, and cover the continental shelves where oil and mineral deposits are beginning to be exploited.

In studying the oceans from space, one is dealing with a dynamic, rather than a static or near-static, phenomenon. Since their dynamism is expressed in terms of short or relatively short time bases—conditions at sea are always changing—permanent observation, monitoring, and acquisition of data are essential. Equally essential is the availability of trained "technical middlemen" who can process and interpret these data and make them available to consumers in readily understandable and usable form.

As oceanographic satellite experience is built up, the predictive value of the data acquired inevitably will increase. Unfortunately, since there is as yet no planetwide information bank of oceanographic processes and actions, one must proceed cautiously to determine experimentally where the strengths and where the weaknesses of satellite techniques lie, and how the former may be applied either to supplement or to replace conventional methods.

A variety of measurements can be made from space by optical, radar, infrared, and other sensing techniques. Principal measurable phenomena include the "heat budget" developed at the ocean-air interface, general circulation of ocean currents, present and predicted sea surface roughness and temperature, and distribution of biological phenomena. As data are accumulated and fed into central processing centers, they are analyzed, checked against whatever "ground truth" comparative information may be available, and then made known to research and commercial users.

BENEFICIARIES OF OCEANOGRAPHIC INFORMATION

Three basic industries are the principal beneficiaries of information derived from oceanographic satellites: (1) the fishing industry, (2) the shipping industry, and (3) a variety of enterprises that for convenience can be grouped and referred to as coastal water industries. They are not, of course, the only users or beneficiaries of oceanographic dividends from space. Dividends also have accrued to government, university, and industrial research organizations in the form of new and better ways of obtaining data for physical, geological, and biological studies of the great oceans of the world, and to meteorologists, who are vitally concerned with ocean phenomena since they strongly affect world weather patterns.

The Fishing Industry

Fishing is one of the world's great industries, and fish is a major and potential source of food for a rapidly expanding global population. Consumption varies from place to place, being relatively low in meat-eating countries like the United States and Argentina and high among such island peoples as the Japanese and the English. Some nations are traditionally fish-oriented, others—notably Peru—have only recently exploited their marine resources. That South American country has, in fact, overtaken Japan as the nation reporting the largest annual catches. The Soviet Union and China do not provide statistical data to the United Nations Food and Agricultural Organization's *Yearbook of Fishery Statistics* or the *Bulletin des Pêches Maritimes* of the International Council for the Exploration of the Sea.

In the two decades since 1950, the worldwide fish catch has increased an estimated 200 per cent, to nearly 65 million metric tons, or well over 140 billion pounds. (One metric ton equals 1.102 short tons equals 2,205 pounds.) About 80 per cent of the fish are found in cold and temperate waters, at depths that seldom exceed 400 to 600 feet. Fish consumption is relatively low in the United States (about 11 pounds edible weight per inhabitant per year, compared with 113 pounds of beef, 65 pounds of pork, 3½ pounds of lamb and mutton, 47 pounds of chicken, and 8 pounds of turkey). Because of consumption rates and strong, low-cost foreign competition, American fishermen find it increasingly difficult to compete in their home market and end up supplying only 40 per cent of the fish consumed in the United States. In 1969, the United States imported 73,694,996 live fish alone, according to the Bureau of Sport Fisheries and Wildlife. Despite low consumption, as a nutrient fish is hard to beat. In comparing a 3-ounce piece of swordfish with a similar weight of roast beef, it is found that the fish supplies 24 grams of protein compared with 17 for the meat, 150 calories versus 375, 5 grams of fat versus 34, and 1,750 international units of vitamin A versus a mere 70 for the beef. (Ironically, this nutritious food was banned in 1971 by the Food and Drug Administration because of mercury pollution; in the future, satellite observations from space may help to monitor such pollution and thus assist in controlling it.)

The problems of the commercial fisherman are to know where fish will likely be found, how large the potential catch will be in a given area, and how rough the sea will be in the area. Oceanographic satellites can help, but only on the basis of predicting, for example, what favorable conditions to look for in the general fishing ground. And a fishing ground, in turn, will be such only if the biotic conditions in the spawning areas that feed it are suitable.

The oceanographic satellite promises to augment and in some cases supplant traditional marine biology research techniques, yielding invaluable information to our understanding of the distribution and habits of life in the sea. Although the principal commercial beneficiary is the fishing industry, increased knowledge of the biology of ocean plants and animals and their interrelationships with the ocean waters and the sediments on the floor may be of fundamental importance as man looks to the sea to nourish his exploding population. The interactions with air and land life of these plants and animals must be learned, as well as their contributions to the ecological balance and their sensitivity to pollutants.

Scientists and engineers have for years worked with biological oceanographers, pondering ways of applying astronautical techniques to marine biology and particularly to the location of fish concentrations. Since fish depend on waterborne nutrients, the study of the biological productivity of the oceans is directly related to commercial fishing—as well as to the growing sports fishing industry. Schooling fish often produce oil slicks and vapors on the surface that may be measurable under some conditions by absorption spectrometry in both the ultraviolet and the infrared regions. Such techniques, while feasible with airborne sensors, may in the long run be less valuable than indirect indicators using orbital observations of the physical environment of the sea and the distribution and abundance of nutrient. Among the observable, and measurable, physical characteristics are surface and subsurface water temperatures; temperature outlines of currents and upwellings; nature and strength of surface winds; and the depth of the water, its salinity, and even its color. Phytoplankton concentrations, the prime food for fish, are heavily dependent on water temperature; hence this physical characteristic can be used to help pinpoint the location of feeding grounds. Regular measurements are needed, however, as plankton concentrations change daily and seasonally; for example, the vertical mixing of water may cause plankton to be drawn below 300 feet, where light intensity is too low to support it.

Simply measuring the temperature of the water or determining the

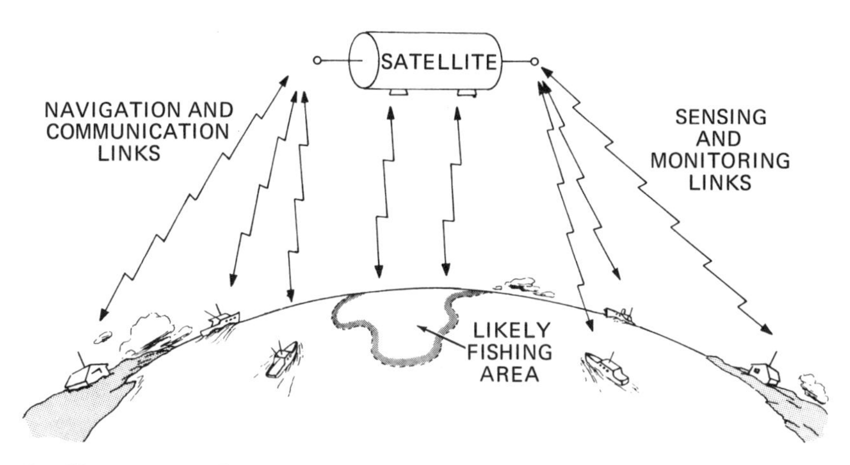

Satellites can greatly improve the efficiency of fishing operations by directing boats to the most promising areas. Sensors may detect oil slicks and surface vapors produced by schooling fish, or may determine the most likely locations of schools on the basis of the physical state of the ocean, the temperatures of interfacing currents, and the abundance and distribution of nutrients. Fishermen obtain the data either directly from the satellite or indirectly via a ground station. (RICHARD P. MCKENNA)

local abundance of plankton is important in inferring the presence of schooling fish, but these two factors of themselves cannot insure success in prediction. One must come to understand the spawning, breeding, migratory, and other habits of each type of pelagic fish, the creature's interaction with the ocean environment, and the ecology of which it is a part. Closely coupled to these factors is the matter of conservation; great care must be exercised not to overfish an area simply because oceanographic satellite technology helps pinpoint otherwise elusive schools. Resource management of marine fisheries must be employed to prevent overharvesting the bounties of the seas.

Satellites may be used to measure, or relay information on, the several environmental and biotic variables. They may determine to within, say, a degree or two of accuracy the sea surface temperatures by infrared sensors, initially in areas where fish concentrations are likely and later on an oceanwide basis. Eventually fishing fleets will receive regular ocean temperature maps of the part of the world where they happen to operate. Actual and predictive maps of sea state will also become available, both of which are of great importance to fishermen.

In order to establish the position of local concentrations of marine plankton, satellites equipped with multi- or bispectral (absorption spectroscopy) instruments to detect surface chlorophyll may be employed. Marine biologists have discovered that small marine animals often concentrate in areas of abundant chlorophyll—at or above a concentration of 0.2 mg per cubic meter. Tuna, for example, frequently assemble in areas where the chlorophyll abundance is above 0.2 mg per cubic meter. Measurements must be accurate to ± 0.1 mg per cubic meter in order to employ chlorophyll as a factor in predicting the location of commercially attractive quantities of fish.

Fishery experts estimate that marine fishing grounds in all the world's oceans should eventually yield an annual crop of some 2 billion metric tons, part of which will come from known and exploited grounds and part from new or partially developed grounds, such as those off India and Somalia. Since the cost is high for establishing a new fishing enterprise or expanding an old one, everything possible should be done to assure the fisherman the best predictive advice possible.

Agencies such as the Bureau of Commercial Fisheries and the Bureau of Sport Fisheries and Wildlife are encouraging the development of astronautical systems to help work out the global fish ecology (including migration cycles, measurement of the worldwide production of phytoplankton, and relation of fish production to thermal currents). It is estimated that the United States spends about $2 million a year alone to gather what little is known about phytoplankton occurrences and concentration. With the advent of better and more economical techniques, fish catches should increase by at least 30 per cent, an increase worth hundreds of millions of dollars a year.

With the advent of satellite oceanography, major progress toward two crucially important objectives appears possible. First of all, fish catches can be made more efficiently and more economically by fewer fishermen using less equipment for shorter periods of time. Even though fish are generally concentrated in rather restricted portions of the world's oceans, the areal extent is very great within the frame of reference of slow-moving fishing craft. If the fleet operator could know where the fish are running and the approximate size of the schools, he would know when to sail, where to go, and how many craft to send. He could economize on manpower at sea, fuel consumption, and wear and tear on his fishing vessels, as well as through the dispatch of the minimum number of craft for the operation at hand and consequent reduction of replacement and repair costs of fishing equipment. It is estimated that about 80 per cent of a fisherman's time is taken up simply in scouting for the fish, compared with 20 per cent in the fishing itself.

The second great benefit to the fishing industry of oceanographic satellites involves the increase in the total number of fish caught due to better exploitation of known fishing grounds, coupled with the development of new grounds. Occasionally fish suddenly leave traditional fishing areas, causing severe economic difficulties until their new grounds have been located.

The potential commercial value to fishermen, to fish processers—including canning and freezing entrepreneurs—to distributors, to transporters, and to the many other entities and individuals involved in making this highly important protein source available to the consumer is immense. Even in the United States, whose fishermen harvest a relatively small percentage of the world catch, the economic impact of the fishing industry exceeds $4 billion a year. (An additional $750 million comes from about a million tons of table fish caught by 8 to 10 million sports fishermen; the value to industry of the purchase of fishing equipment, clothes, boats, and the like, is not considered.)

The short- and long-range benefits of satellite oceanography applied to the world fishing industry should not be underestimated. All evidence points to the steadily increasing cost of such "luxury" protein foods as beef and lamb, partly as a result of the rising value of the land upon which the animals feed and partly as a result of human wage costs at all levels of the handling, packaging, distribution, and sales processes. The day may come sooner than one expects when steaks and chops will be priced beyond the daily reach of the average consumer, who may be forced to seek his protein from the rich harvests of the seas. Oceanographic satellites may well prove to be the tool to keep this vital food within his reach. A University of Rhode Island study of the tuna fishing industry showed that, if a satellite system could reduce the time spent searching for fish by 50 per cent, the per-day catch would increase by 25 per cent, resulting in not only an annual saving of $15 million but a decrease of $7.5 million in equipment investment. Such savings presumably would be reflected in the price of fish.

The Shipping Industry

It is estimated by the Department of Commerce that the United States alone pays in excess of $50 billion a year for imports and exports carried by sea. The shipping bill, represented by approximately 400 million metric tons of cargo, amounts to about $5 billion, to which must be added another $5 billion in related costs. Seen from the point of view of an individual shipowner, operating costs run from $1,000 to $4,000 a day. With the continued growth of the world merchant fleet, already exceeding 200 million tons, and the advent of such superships as the Japanese-built 312,000-ton tanker *Universe Ireland*, mankind's investment in sea transport is certain to soar—as will his desire to control costs by all possible means.

The application of oceanographic satellite techniques to ocean transport and coastwise shipping (and, logically, to the movement of pleasure, fishing, and naval vessels) promises to be a major factor in improving both routing and safety. From a commercial point of view, about half of the shipping bill represents costs incurred while a freighter or tanker is on the high seas. If the time at sea can be reduced—and the voyage made safer for the ship itself and its crew and cargo—consequent savings (on the order of half a billion dollars a year for the United States) should be realizable.

Once oceanographic satellites are fully operational, they will be able to provide synoptic coverage on a regular basis of the state of the oceans all over the world, a knowledge with numerous advantages. For example, ships will be informed of, and thus be able to shun, routes along which seas are particularly rough. By avoiding areas of high sea state (and severe weather and adverse currents), travel time can be reduced. During a year-long test program with conventional, nonsatellite routing aids, the Military Sea Transportation Service of the United States saved $4.5 million in direct and $12 million in indirect costs with 1,600 cooperating ships. Advantages cited were: (1) reduced travel time, (2) reduced fuel consumption, (3) reduced

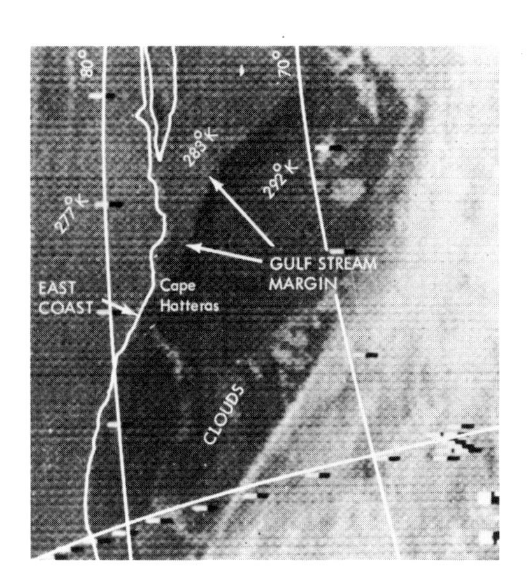

High-resolution infrared imagery from the Nimbus 2 satellite clearly shows the Gulf Stream boundary. Such information permits ship captains to route their vessels so as to take maximum advantage of favorable current flows and avoid adverse ones. (NASA)

general operating costs, (4) reduced cargo damage and loss, (5) reduced ship damage, (6) increased passenger and crew comfort, (7) reduced insurance rates, and (8) better ship scheduling. Concerning the last, stevedore waiting time can be minimized at port, along with standby dock costs committed to a tardy ship. Moreover, cargo pickup arrangements can be more efficiently planned, including cargoes of opportunity, and ship turn-around time cut down.

In order to improve their routes, ship captains and navigators plying the world's oceans need prognostic weather charts, synoptic and prognostic wave charts, temperature readings, and data on major currents, all regularly upgraded by inputs from orbiting oceanographic satellites. For example, by determining current boundaries, ships can be routed so as to take maximum advantage of flows in favorable directions and to avoid them when movement is contrary to the destination. Without such satellites, only between 500 and 1,000 sea-state reports are supplied on a given day from surface ships; and these are clustered along major shipping routes, are mostly visual, and are not overly reliable simply because they usually are made only during favorable conditions. With the prospect of accurate short-, medium-, and long-term ocean condition forecasts from satellites, all shipping interests—commercial, government, and pleasure alike—will benefit.

Other than rough seas, bad weather, and currents running counter to the direction of travel (the Gulf Stream moves at from 2 to 5 knots), vessels must contend with fog, sea ice, icebergs, derelicts, and shoals—all navigational hazards detectable by sensors. Although the International Ice Patrol has greatly reduced the danger of ships colliding with icebergs since the *Titanic* disaster that claimed 1,517 lives, it is costly and not wholly effective (in 1959, the *Hans Hedtoft* went down off Greenland with 95 aboard after hitting an iceberg). The patrol maintains both surface and aerial surveillance of potentially dangerous areas, augmented by national efforts of major interested maritime powers. The United States and Canada spend about $10 million a year on aerial ice reconnaissance, figured at the rate of about $500 per flight hour; and Canada alone recently spent $5 million just to instrument two ships for ice surveying. Ice watching is an expensive enterprise. From the shipowner's point of view, icebergs not only pose the prospect of disaster but also cause navigators to guide ships along circuitous routes at slow speeds simply because their locations are unknown.

The occurrence of sea ice and icebergs is of vast commercial, military, and scientific importance. Basically, the users of information relating to ice need to know its present, immediate future, and probable longer-term (that is, within the month) state. Among the many observables are sea-ice boundaries, percentage of ice concentration per square unit of measurement (for example, per square mile), topography, thickness, and tonal variations. Others include type and age of ice; nature of meltwater puddles, cracks, and other openings; and drift direction and rate. In addition, there are time and rate of breakup of coastal ice in the spring season and time and rate of closure as autumn and winter progress.

Satellites equipped with all-weather sensors offer solutions to many complex ice and ice-related problems beyond detecting, monitoring, and predicting the paths of icebergs. Information is also needed on the birth and population of icebergs, their form and thickness, and the winds and currents that affect their movements. Similar data are required on sea ice (and to an extent on river ice as it moves out to sea), particularly in the light of the growing movement of ships in the Antarctic, through the Northwest Passage and Hudson Bay, around Greenland, and through the Northeast (Siberian) Passage and other parts of the Arctic. The Russians require continuously updated prognoses of routes leading to their expanding northern ports and of the likelihood of their being closed by ice.

The commercial importance of both acquiring basic knowledge about sea ice and providing regular surveillance of it is underscored by the discovery of an immense oil field at Prudhoe Bay, Alaska, whose reserves of low-sulfur crude oil have been estimated at from 5 to 30 billion barrels. To help assess the feasibility of moving some 2 million barrels of crude oil a day to world refineries from Alaska's north slope coastal plain, new icebreakers have been and are being built; and between September and November 1969 Humble Oil and Refining Company's 150,000-ton tanker S.S. *Manhattan* made an exploratory round-trip passage through pack ice across the Northwest Passage, the first commercial ship to do so.

The cost of construction of Prudhoe Bay docking and loading facilities may run as high as $500 million, indicative of the economic stakes involved. According to British Petroleum's Captain Ralph Maybourne, it "will be the most difficult accomplishment of its kind in history." Once built, the port would have to be regularly reachable through the Northwest Passage, the feasibility of which was demonstrated by the *Manhattan*'s Arctic odyssey.

The 10,000-mile trip was successfully made, although the tanker had to be freed from enclosing ice at least a dozen times by the *John A. MacDonald*, a 315-foot-long Canadian icebreaker displacing more than 8,000 tons. Despite damage incurred in the ice, it seemed probable that the *Manhattan*, and other ships, could make the trip again. Uncertain, however, was whether commercial service could be maintained on a year-round basis. If so, per-barrel costs to the East Coast of the United States would be about $1.00, between 35 and 45 cents cheaper than using an alternative transcontinental pipeline across Alaska and Canada. If the Northwest Passage proved feasible from a commercial point of view, it would likely soon be used for more than shipping oil. Japan and Europe would immediately become 8,000 miles closer to each other, a fact of vast trade portent.

Even if the Northwest Passage cannot be maintained open all year, it must inevitably become a trade route part of the time, especially when larger and more powerful icebreakers become available. As for the Northeast Passage, Russia reported in June 1968 that it had several atomic icebreakers under construction to keep the Arctic sea lanes open for longer periods during the year than now possible. As more ships travel into zones of heavy sea ice, reliance on ice-surveillance satellites will increase. Four ice-forecasting services are envisioned: (1) long-range, on the order of 3 months; (2) medium-range, about 1 month; (3) short-range, about 5 days; and (4) daily. In addition, oceanographic satellites could observe, and report on, ships in distress, and help to guide rescue efforts. And, of course, they could also maintain an inventory of vessels in polar waters—and elsewhere around the world as well—either by direct optical observation or, more logically, by locating and interrogating ships with transponders mounted on them.

Ice surveillance has other beneficiaries than shipping. Meteorologists, for example, need continuous inputs to help them make valid weather predictions and climatic models. Oceanographers observe the movement of icebergs as indicators of the direction and speed of ocean currents; hydrologists use ice data in calculating the amount of water tied up in such inland bodies as bays, sounds, and lakes; and marine biologists take ice into consideration as they study biological productivity in far northern and southern waters.

American Tiros and Nimbus satellites and their Soviet counterparts have proved the feasibility of ice surveillance over both oceans and large lakes and have revealed that the tracking of loose sea ice and icebergs has the double benefit of locating and following the ice masses themselves and of determining the velocity and direction of the ocean currents that carry them. These pioneering craft were not designed primarily to study the oceans, but they do point the way toward future oceanographic satellite systems and provide test beds for both active and passive sensing devices.

Coastal Water Industries

Closest to the land, and inextricably tied to it from the point of view of physical, environmental, and human activities, is the coastal region. Included within the coastline boundary are estuaries, bays, inlets, harbors, and, of course, thousands upon thousands of miles of sandy beaches. It is at the coastline that the oceans react with the land; that hurricanes, tides, currents, and waves either erode material from or accrete it to the shores; and that ships arrive and depart with freight and passengers. It is here, too, that often dangerous shoals pose hazards to shipping, and that oil and mineral wealth is increasingly exploited. Also, swimming, scuba diving, surfing, boating, sport fishing, and other recreational activities take place here. Rivers empty their water and sedimentary cargoes here—together with, unfortunately, thermal, biological, chemical, and mineral pollutants. While it is impossible to estimate accurately the tangible and intangible value of space-derived observational data on the coastline, it exceeds that of observational data on the more commercially identifiable fishing industry. Someday it may be possible to assign dollar values to clean estuaries and beaches on the one hand and coastline–continental-shelf mineral wealth on the other.

Synoptic observations of the world's shorelines can be made by photography, a technique successfully demonstrated by such unmanned satellites as Nimbus and Dodge as well as by manned Mercury, Gemini, and Apollo spacecraft. Video telemetry links and microwave radar may be useful for some applications, but their outputs do not compare in quality with photographs taken by telephoto-lens cameras. Routine surveys will normally suffice for shorelines in general, with more detailed observations being reserved for special targets such as important rivers, harbors, estuaries, and heavily built-up recreational areas. The last of these targets is far from insignificant, in view of the facts that (1) fully a third of the world's rapidly expanding population lives near the shore, (2) the beaches and other shoreline recreational sites are confined in area, and (3) more and more people are spending their holidays on the coast.

The principal beneficiaries of coastal oceanographic information can be grouped conveniently into seven categories, most of which are closely interrelated. They include, in addition to the estuarine and coastal fishing industries already considered, (1) port and harbor, marine transport, and coastwise shipping interests; (2) coastal engineering and submarine topographic surveys; (3) continental-shelf industries; (4) recreation, including sports industries; (5) sea-danger warning organizations and systems; and (6) agencies and authorities involved in water quality and pollution control.

Transport. Oceanographic satellites can aid in maintaining inventories of vessels converging on and departing from channels of heavy marine traffic. They can also assist in surveying the feasibility of building port facilities near discoveries of new natural resources, such as those at Prudhoe Bay, Alaska (oil), and along the Gulf of Carpentaria, Australia (bauxite).

Surveys. Coastal engineers and submarine topographers must gather, on a regular basis, data needed by port and harbor authorities and other shoreline groups. This information includes erosion rates of beaches due to normal wave action, currents, tides, and severe storms; sedimentation rates of estuaries and harbors by rivers; and regular surveys of shoal and shallow water. Using suitably filtered color photographs, engineers can prepare shallow-water depth contours for areas where color contrasts are high. These often reveal shoals, bars, reefs, and derelicts, which can then be accurately mapped. Even shallow seamounts may be observed by their effects on ocean temperatures and circulation configurations. The resulting improved nautical and hydrographic charts will be of inestimable benefit in safety, economy, and efficiency of marine navigation.

The charting of shoals and reefs by conventional techniques is expensive, time-consuming, and often inadequate simply because of the vast areas involved. However, it is feasible to provide continuous oceanic surveys by polar-orbiting satellites using both color photography to detect clear-water shoals and infrared imaging systems to reveal turbid-water shoals. In principle, once a thorough satellite survey has been made of the oceans, it should be possible to chart all dangerous shoals. However, since shoals undergo continuous morphological changes, repeat reconnaissance will be required.

Continental Shelf. The exploitation of the continental shelf during the past few decades has been spectacular, to the profit of innumerable coastal industries. The most important of these, the petroleum industry, estimates that 16 per cent of world oil now comes from continental shelves and that some 50 nations are either exploiting them or making ready to do so. In accordance with the International Convention on the Continental Shelf, signed in Geneva in April 1958, maritime nations secured title to the offshore shelf out to the 200-meter contour line—or, much more vaguely, "to where the depth of the superjacent waters admits to the exploitation of the natural resources." In the case of the United States, as well as some other countries, this resulted in valid claims to enormous oil reserves. The

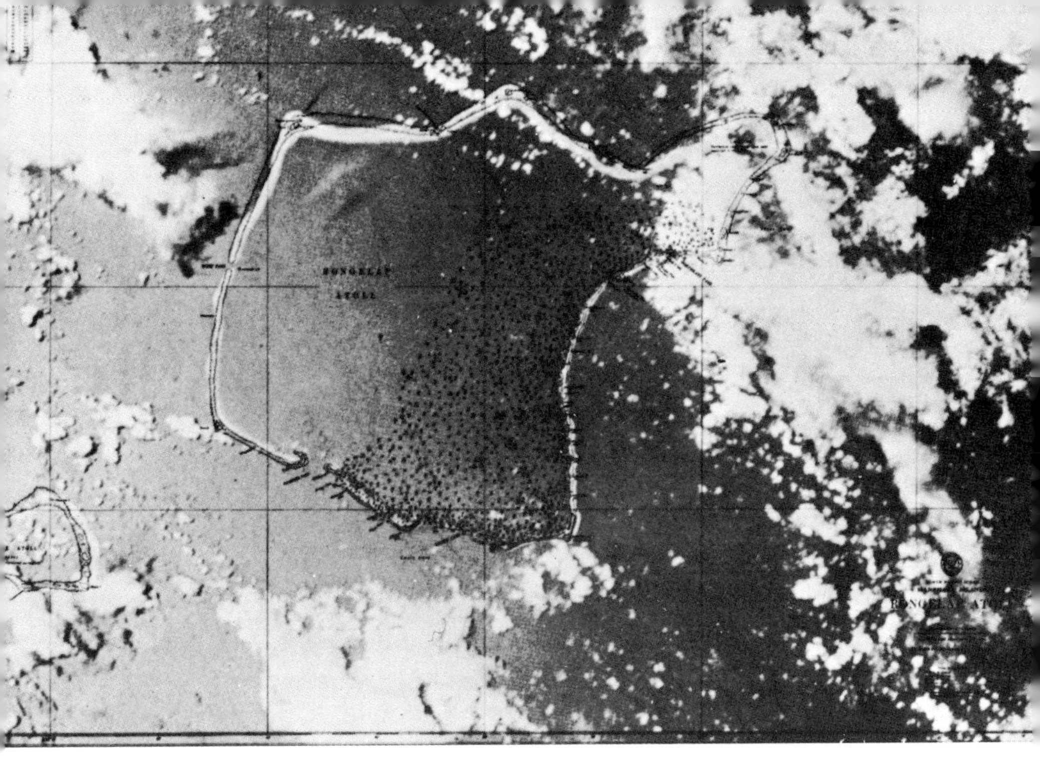

Oceanic surveys by satellites are providing mapmakers with a wealth of new information, enabling them to correct old errors, to fill in more details, and to update their charts by taking into account the ever-changing shapes of reefs and shoals. Superimposition of a navigation chart on this Gemini 5 photograph revealed positional and configurational errors in the northern and northwestern sections of Rongelap Atoll in the Pacific. (NATIONAL COUNCIL ON MARINE RESOURCES AND ENGINEERING DEVELOPMENT, AND U.S. BUREAU OF COMMERCIAL FISHERIES)

annual value of marine minerals extracted from the United States continental shelf runs about $100 million, and is going up every year. To help foster coastal industries, satellites can photograph the shelf, interrogate buoys, measure waves and thermal conditions, possibly detect spectroscopically vapors related to underwater mineral deposits, and contribute to the worldwide Continental Shelf Mapping and Charting Program. In order to assist in the latter, which includes the extension of geodetic controls from land out onto the shelf, oceanographic satellites may be used to pinpoint the location of ships anchored over various bathymetric contours, say, the 100-, 150-, and 200-meter depths. By orbital photography, these ships could be accurately tied into the land-based geodetic grid.

Recreation and Fishing. The coastline and the waters adjacent to it are the source of great pleasure to peoples all over the world, and major industries have emerged to both enhance and exploit it. In the United States, with nearly 22,000 miles of coastline (of which somewhat over 1,200 miles are open to public facilities), the coastal recreation industry brings in about $2 billion a year, and is increasing at the rate of $100 million a year. The sites for recreation include not only beaches but bays, inlets, sounds, inland waterways, and scuba-diving areas. In these same waters, sport and commercial fish are caught and shellfish harvested.

Warning Systems. Port, harbor, continental-shelf, coastal engineering, recreation, and fishing industries are all susceptible to storm surge and tsunami (seismic sea wave) damage. Sea-danger warning systems can take advantage of oceanographic satellites to detect these devastators of life and property. It may even be possible for underwater volcanic eruptions to be detected by measuring thermal gradients with infrared line-scan imaging systems, by measuring escaping gases by absorption spectroscopy, and by photographing the changing color of the water caused by transport to the surface of materials from the sea bottom.

Oceanographic and meteorological sensors can combine to monitor the buildup of typhoons, cyclones, and hurricanes and the storm surges, or abnormal rises of coastal seas, they produce. Radar and laser instruments may detect potentially dangerous ocean-level changes rather far out at sea, and help to predict just how destructive they may be to man and his works. Among the variables that are to be considered are the distance the storm surge must travel, the size of the storm and the speed of its winds, and the topography of the ocean bed over which it travels. Hurricane Camille, which struck the Gulf Coast of the United States, and a monstrous Pacific storm 1,500 miles northwest of Hawaii caused great loss in life and property during 1969. Experts estimate that losses would have been far greater if the storms had not been tracked by weather satellites (see chapter 7) and warnings issued based on their observations. Up to 50,000 persons might have died in Camille alone had satellite data not been available. The terrible autumn 1970 East Pakistan typhoon was observed by weather satellites, and advisories were made available, but ground warning systems were ineffectual, and hundreds of thousands of persons lost their lives by drowning. In earlier years, storm

While weather satellites warn of the advance of hurricanes, reconnaissance satellites provide valuable service in assessing damage after they have hit. Hurricane Beulah in 1967 made dramatic changes in the Texas coast, evidenced in these before (left) and after (right) photographs. (U.S. GEOLOGICAL SURVEY)

surges were produced along the Atlantic coast between Savannah, Georgia, and Charleston, South Carolina, in 1893, killing well over 1,000 people; and along the Texas coast near and at Galveston, taking 6,000 lives in 1900. Storm surges only a few feet high can flood large areas of low-lying land, causing much damage and loss of life.

Tsunamis can be extremely destructive. Hawaii, for example, has been hit by 85 since 1814, when records on them were first kept. Moving as fast as 600 mph, they may be recorded by surface, aerial, and satellite sensors that detect changes in sea level and sea slope and altered eddy patterns over submarine ridges and seamounts. The same satellites that help warn of impending storm surges and tsunamis can also perform coastal damage surveys after they have hit, and help rescue teams to assess where best to direct their efforts. In the 1969 Camille aftermath, many cut-off Mississippi communities could not communicate their pleas for assistance, but satellite pictures clearly showed the extent of their plight.

Satellites can, in addition, chart changes in natural coastline features, such as reefs, offshore bars, and harbors.

Pollution Control. Partly of practical and partly of esthetic importance is the matter of coastal water purity, which is being severely compromised by the horrible specter of pollution. In the United States, the situation became so serious that the Federal Water Quality Administration undertook studies of estuarine pollution to seek out effective means of halting the growing hazards to human health, to recreation, to shellfish beds, to fisheries, and to general ocean ecology. Once again, satellite sensors can assist in measuring water properties, including the buildup of pollutants, by observing the boundaries of effluents (which may indicate stagnation in estuary inputs from polluted rivers, and thermal pollution). They also can detect color clues (which can reveal chemical pollution, polluted fresh-water inflow, and pollution caused by nuisance algal and weed growth) as well as unusual chlorophyll content (indicative of some chemical and biological pollution). Because of the vast geographical areas involved and the ever-changing nature and concentration of pollution, satellites offer the only feasible system of making repeated, regular worldwide surveillance of the ever-growing danger to mankind.

Pollution of the coastal regions results from man's misuse of them as dumps for his wastes: trash, raw sewage, and industrial refuse, including chemical sludges, slaughterhouse debris, metal foundry byproducts, military warfare chemicals, oil spillage from tankers and refineries, dumped materials from dredges, and all manner of agricultural chemicals. The U.S. Public Health Service is almost certain that the pesticide endrin was the agent that destroyed 10 million fish in the Mississippi River–Gulf of Mexico brackish water zone in 1964.

Industrial enterprises and local municipalities realize the difficulty but do not have adequate funds to do much about it. One observer pointed out that "public enthusiasm for pollution control is matched by reluctance to pay even a modest share of the cost." Such an attitude will have to be changed radically in the very near future if the coastal regions, and even the oceans themselves, are to be preserved from man's carelessness.

Despite concern and some good intentions, pollution continues unabated. Long Island Sound was twice as polluted in 1970 as a mere decade earlier. Shellfish harvesting has been tragically affected, as have the sound's esthetic and recreational qualities. Many estuaries are unfit and more are becoming so, producing millions of acres of hazardous shellfishing grounds, among other things. Realizing that the problem is international in scope, the Intergovernmental Maritime Consultative Organization of the United Nations is seeking a way to develop laws on coastal pollution.

WHAT CAN BE LEARNED ABOUT THE OCEANS FROM SPACE?

Many vital aspects of oceanography can be aided by satellite technology; indeed, this important new tool promises to revolutionize the study of the seas, augmenting conventional research techniques in some cases and entirely replacing them in others.

First, oceanographers can obtain from remote sensing and data relay satellites ocean and sea data on a synoptic, global scale, leading to an increased understanding of the broad dynamics of marine processes. This can be obtained by repeated measurements of the physics and chemistry of water, the actual and predicted state of the sea, currents and their boundaries, anomalies expressed by the physical appearance and topography of surface waters, the topography of shorelines, the nature of continental shelves and offshore shoals, and the character of the vast and poorly understood sea floor. Satellites can also provide information on the contribution of the oceans to the planetary environment, including the global heat budget. More detailed, localized data on how the ocean influences the weather are being sought—for example, changing circulation patterns (currents, upwellings, and sinkings).

In order to increase man's understanding of the oceans and apply it to sea transportation, fishing, and coastal water industries, much specific information is needed—and needed on a regular, and global, basis. This information varies from the nature of the instantaneous and predicted state of the sea in a given area, to the composition of the water, to the whereabouts and probable drift rate and direction of icebergs.

TABLE I
OCEANOGRAPHIC PHENOMENA MEASURABLE FROM SPACE

MEASURED CHARACTERISTICS	TYPES OF SENSORS AND PROCESSES
Sea state	Radar roughness scatterometer (radar backscatter for wave-height measurement); passive microwave radiometry
	Cameras for sun glint and cloud patterns
Thermal contours; absolute sea temperature	Infrared imagery; radiometry; multiple-frequency microwave radiometry; spectrometry
"Topography" of sea surface; seismic sea waves	Radar and laser altimetry
Sea ice, icebergs	Cameras (panoramic, ultrahigh resolution); multiband synoptic photography; radar imagery; infrared imagery
Currents	Infrared imagery; microwave imagery; multiband synoptic cameras; interrogation of floating platforms
Composition of water	Infrared imagery; cameras; spectrophotometry
Coastal mapping; shoal waters	Cameras; radar; infrared imagery
Biological phenomena	Optical observation (sea color); multi- or bispectral absorption photography (bioluminescence, fluorescence); infrared radiometry
Surface and subsurface phenomena	Buoy-to-satellite data relays (all instruments on buoy platforms); Interrogation, Recording, and Location System and Omega Position Location System
Water vapor [a]	Passive microwave radiometry and imagery

[a] In atmospheric column between sensed ocean phenomenon and satellite sensor.

Sea State

The force of the wind blowing across the surface of the ocean is the major factor governing its "state" at a given place and time. This state, or condition of wave heights and patterns, is commonly known as the roughness of the surface. It is represented by a number, which varies approximately with the square of the velocity of the wind. Most of the momentum transferred to the oceans from winds goes into making waves, which gradually dissipate their energy by breaking; waves carry both energy and momentum.

Conventional sea-state information is derived almost exclusively from coastal regions and along the major shipping lanes by ship-mounted and buoy-mounted wave recorders, and occasionally from aircraft observations with radar and laser altimeter wave-profile recorders and stereo cameras. These methods are acceptable for local coverage; but, because of the restricted area they can cover, they are inapplicable in generating synoptic data of global sea state.

The advent of the satellite has made available a tool of enormous potential benefit to oceanographers. A technique involving the use of a radar scatterometer has been developed that can detect the gross characteristics of ocean roughness by comparing a number of radar return signals. In calm seas the radar return is essentially vertical, but as roughness builds up, reflections move away from the vertical because of scattering from the waves. By analyzing the angular distribution of the radar return, the radar roughness can be worked out, which is related to the visual sea state.

Another method involves a device known as a multispectral microwave radiometer, used to measure microwave emissions that are correlated with the state of the sea. Other devices, such as the synthetic aperture, side-looking radar, and several laser systems, can also yield sea-state profiles. Also, experimental work with Tiros, Essa, and other satellites has been undertaken to relate solar glint effects and the nature of the sea. As ocean roughness increases, regular, nondiffused ("specular") reflections of the sides of the waves cause a halo to be formed in photographs. Preliminary research shows that the larger the halo, the more active is the sea below.

Sea Level, Tides, and Slope

Sea level, often referred to as dynamic topography, is conveniently thought of as the average height of open, or free, ocean water halfway between the crests and troughs of the waves. It is shaped by currents, tides, the atmosphere (winds and pressure), and the rotation of the planet, and as such is ever-changing.

Only spotty data on sea-level anomalies are now available, but satellite techniques hold promise of making available information on a large-scale basis. The maximum vertical drop between ocean "valleys" and "hills" is of the order of 5 feet. The situation is similar with regard to information on ocean slopes (which vary in steepness from 1 vertical unit in 10,000 horizontal units to 1 in 100,000, depending on the velocity of the currents that cause them) and open ocean tides (which probably do not vary more than 10 to 15 inches).

Sea tides, levels, and slopes could be measured by submerged stationary buoys with floating antennas, which in turn could be interrogated regularly by satellites. The satellites would then read out the acquired data to ground or ship stations upon command (or at preset intervals), providing global information on a regular basis. Satellites in equatorial, stationary orbits could conveniently interrogate buoy platforms in low and middle latitudes, while polar or highly inclined orbits would be best for high-latitude readout.

Active radar altimeters, carried in 250- to 300-mile-high inclined- or polar-orbit satellites, may also be able to provide a time-averaged level of the oceans. Precisions of ±5 to ±10 cm would be required to make satellite altimetry useful for oceanographic applications.

The use of laser altimetry also is being investigated.

Surface and Subsurface Temperatures

It is feasible to employ artificial satellites in two fundamental ways to obtain the surface temperatures of the oceans: (1) by direct infrared radiometer measurements from polar or highly inclined orbits, and (2) by satellite interrogation of buoy platforms with temperature-measuring sensors. In the latter case, thermal profiles of the subsurface as well could be obtained by using submerged sensors, whose output could be transferred through the buoy staff to radio-link antennas.

Alternative methods for obtaining sea temperature data on a global scale are not promising. If temperature sensors and transponders were to be installed on large numbers of freighters, data points would be concentrated along major shipping routes, leaving much of the world uncovered; installation and maintenance costs would be high, and outputs would be degraded by both inherent navigational errors and reliance on essentially untrained observers. If only oceanographic research ships were to be employed, the time intervals between readings at different points would be unacceptable to give synoptic thermal profiles on an oceanwide basis. As an illustration of the difficulties involved in using ships, if a single vessel were to cover 10,000 square miles of surface per day, it would take 37 years to sample the oceans of the world.

Since the temperatures of the surface and outer layers of ocean waters are of great importance to the understanding not only of the ocean itself but of the lower atmosphere that interacts with it, the establishment of synoptic or quasisynoptic horizontal temperature profiles is a major goal of both oceanographers and meteorologists. Such knowledge can help man to understand how energy is exchanged between the ocean and the atmosphere, how this energy affects the Earth's heat budget, and how circulation patterns are influenced. As benefits, it is possible to predict climatic and local weather changes, to understand local and regional biological environments and productivity, including fish distribution, and to detect and monitor estuary effluents and pollution.

Initial infrared radiometry systems flown at polar-orbit altitudes of some 250 miles would be capable of making day and night measurements of sea temperatures over cloud-free areas with an accuracy of about a degree centigrade. The analysis of the data must take into account the fact that the infrared readings can be degraded by the presence of a "column" of water vapor between the point on the ocean being measured and the satellite sensor. If oceanwide surface temperature contour maps are to be prepared, it is necessary to use information gained from surface sensors so that contours can be accurately extrapolated under the cloud-covered areas.

Circulation Patterns

Circulation patterns, or gross movements of ocean waters, have been observed for years but are still understood only to a limited extent. It is known that ocean circulation is generated principally by the winds; and, conversely, that the atmospheric "heat engine" is influenced to a great extent by the oceans. The Earth—land, oceans, and atmosphere—enjoys a heat balance, an equilibrium condition existing between the radiation received from the Sun and that emitted by our planet's ground, water, and air surfaces. The prime driver of atmospheric processes is ultimately the Sun.

Just how a given section of the ocean will respond to winds at a given time depends on its latitude, the strength and duration of the winds, the depth of the sea, the nature of the floor, and the proximity of continental or island barriers. North Atlantic and Central Atlantic waters, for example, move in response to prevailing northeast and southwest winds, respectively. However, the resulting ocean currents vary with time, in intensity, and to an extent geographically, contributing positively or adversely to ocean transport (ship captains like to "ride" favorable currents and avoid unfavorable currents) and to the fishing industry (an annual meander of a warm water current causes fish to leave usual feeding grounds and seek others). Unfortunately, conventional oceanographic exploration methods permit only limited aerial coverage within a given period of time, making oceanwide data almost impossible to attain. Nor are these same techniques conducive to sustained surveillance of local regions to make accurate analyses of transitory changes; individual ships simply cannot remain in the same place for months at a time nor visit hundreds of thousands of buoys scattered around the world.

Among the principal mass movements of ocean waters are constant, known currents such as the Gulf Stream and Humboldt Current; meandering, intermittent, and still unknown currents; interactions between different currents; upwellings and sinkings of water masses; and the effluents of estuaries. Tests with a number of satellites show that many aspects of ocean circulation can be observed, monitored, and subsequently analyzed with spaceborne instrumentation. Photography is sensitive to variations in water color, indicative of current patterns, while infrared radiometers detect changes in the thermal conditions of the water—another indicator. Optical, infrared, and microwave radar tracking of objects (including free-floating buoys) drifting with major currents is another valuable technique. Occasionally, the observation of certain cloud structures (characteristic of ocean current boundaries) proves useful.

Observations from orbit of marine biologic phenomena can be applied to increasing our understanding of the dynamic processes of the oceans as well as purely biological conditions. For example, by surveying the drift movements of floating seaweeds, it should be possible to shed further light on circulatory patterns of the oceans. Taking advantage of the fact that during periods of calm seas the exposed surfaces of seaweed are dry or nearly dry, and therefore, like other broadleaf vegetation, highly reflective in the infrared, cameras using infrared color film could be used to produce high-contrast pictures—the seaweed showing up bright red and the water blue-green.

Such observations can increase our understanding of the dynamic forces that lead to the accumulation of millions of tons of floating sargassum weed and the marine creatures associated with it in the Sargasso Sea. Almost without winds and strong flows of water, the vast sea (which occupies an area about the size of the United States) is encircled by the very ocean currents that produced it. Apparently the source of the seaweed is brown algae removed from the Caribbean coasts and islands and carried off by the Gulf Stream, which eventually discards them in the midocean "sea."

Water Composition

On the average, each 1,000 parts by weight of seawater consists of 965.1 parts by weight of water itself and 34.9 parts by weight of salt. Virtually all of the salt is made up of sodium, potassium, magnesium, calcium, and strontium, together with sulfates, chlorides, bromides, fluorides, and carbonic and boric acids. The oceans also contain the plant nutrients phosphorus and nitrogen; trace elements like bromine, iodine, vanadium, and gold; biologic matter; and various dissolved gases. It has been estimated that in a cubic mile of ocean water there are 166 million tons of dissolved salts.

Certain gross variations in water composition can be detected remotely through variable spectral absorption, backscattering, and reflectance of visible light. Variations in film images photographed from space should reveal the boundaries between waters of slightly varying composition, particularly with regard to biologic matter. Thus, infrared color photography could detect the so-called "red tides," resulting from a typical zooplankton concentration, which often cause the death of large quantities of fish by the production of toxic materials.

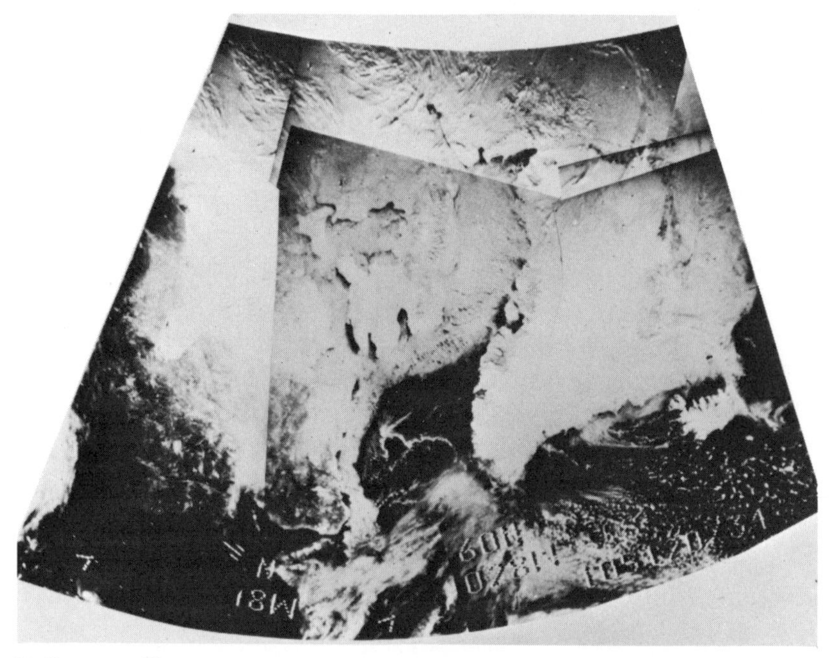

Daily surveillance of the dispersion of sea ice is of great value to the shipping industry. This mosaic of Nimbus 3 photographs, taken from an altitude of about 700 miles in three successive orbits on 15 April 1969, reveals that ice along the eastern coast of Greenland (right center) is breaking up. Labrador coast and the open St. Lawrence Gulf (lower left) stand out against the snow-covered landmass. (NASA)

Sea Ice and Icebergs

Of all oceanographic phenomena, sea ice has so far proved the most readily measurable. This is fortunate, as sea ice's "habitat" is in the far southern and northern regions of the world where climatic conditions make accessibility and observation difficult and often impossible.

Antarctica and its contiguous seas in the south, and Greenland and the northern polar seas produce icebergs and sea ice. These move northward and southward, respectively, the distance traveled depending on their thicknesses, ocean currents, water temperatures, and air temperatures. Southern polar shelf ice commences to form toward the end of January, and by March the majority of bays and inlets are frozen over. Breakup occurs during the spring months; and, de-

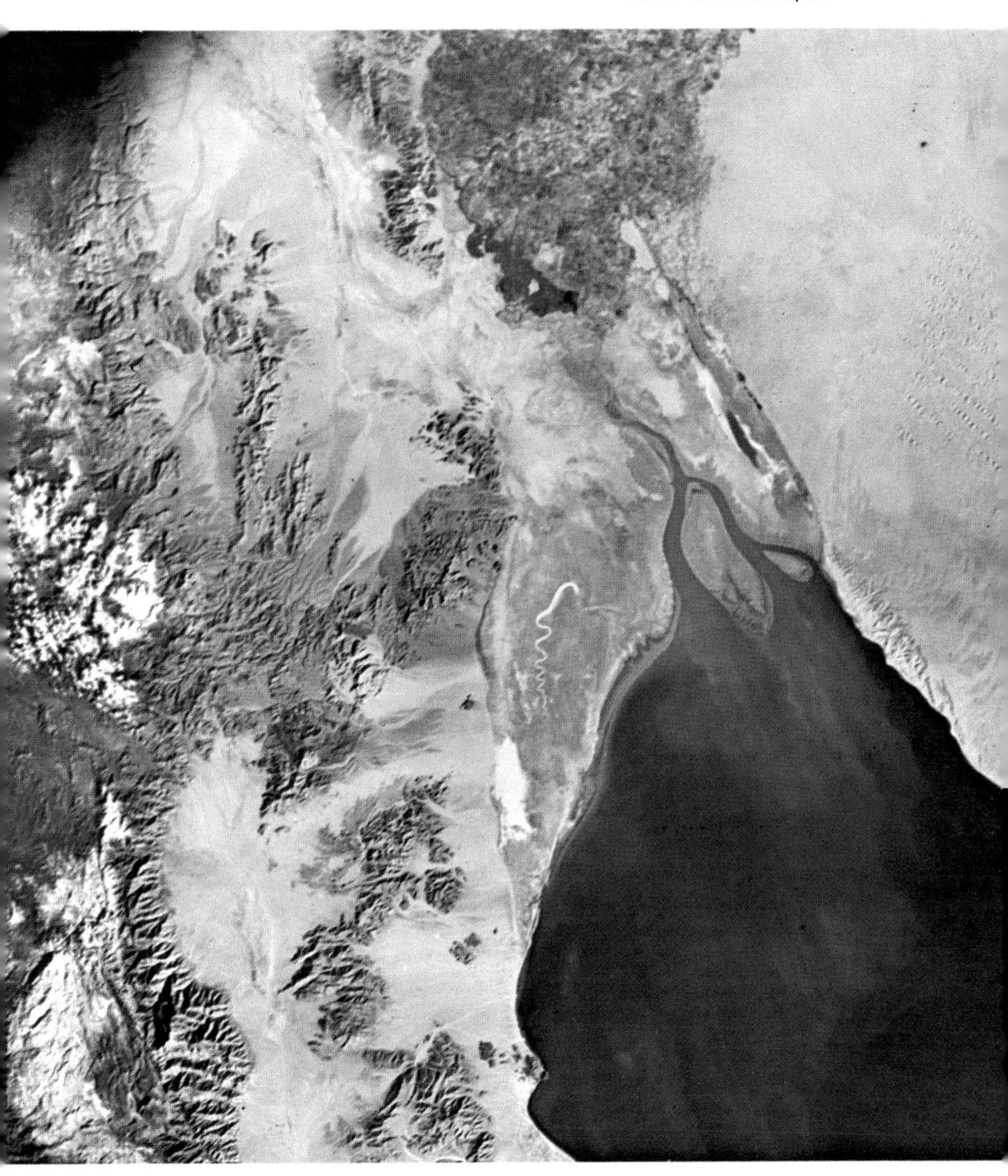

Irrigated farmlands near mouth of the Colorado River stand out in red checkerboard pattern in this infrared photograph, taken during the Apollo 9 mission in 1969. Light colors of silt bands in waters of the Gulf of California reflect currents and depths. (NASA)

Agricultural diseases and pest attacks can be sensed remotely by infrared photography before they can be detected from the ground. Here, insect-infected trees in Oregon appear blue-green, while healthy trees appear red or pink. (NASA)

TABLE II
CONTRIBUTIONS OF OCEANOGRAPHIC MEASUREMENTS TO MARINE TECHNOLOGY OBJECTIVES

MEASURED CHARACTERISTICS	OBJECTIVE SERVED (percentage of use)			
	FORECASTING	COMMERCE	FISHING	TRANSPORTATION
Sea temperature	40%	8%	16%	n.a.
Sea state; wind speed	28	55	18	43%
Current direction	8	9	3	16
Current speed	6	8	3	18
Wind direction	4	8	2	7
Sea surface oil slicks	n.a.	9	30	n.a.
Sea ice	n.a.	n.a.	n.a.	13
Sea color	n.a.	3	6	3
Fish sightings	n.a.	n.a.	22	n.a.
Cloud cover	14	n.a.	n.a.	n.a.

Adapted from an aerospace/oceanographic study prepared in 1967 by the National Council on Marine Resources and Engineering Development and General Electric Company.
n.a. = not applicable.

pending on the depth of the ice, local wind conditions, and geographical protection afforded to individual bays or inlets, it will move northward during December, January, and February. Much of the pack ice remains in a belt around the continent; but some sea ice and individual, large icebergs travel considerable distances, occasionally posing a danger to shipping.

In the Arctic, sea ice covers virtually all of the ocean during the winter. In spring and summer, however, it breaks up, giving rise to areas of open water. The pack ice drifts clockwise around a center geographically located in the middle of the Arctic Ocean. Also found in the ocean are giant icebergs that break off from glaciers, principally in Greenland, Franz Josef Land, and Spitsbergen. Added to them is an occasional ice island, which may be 15 or more miles long and as much as 200 feet thick. During the months of May and June, icebergs off the Grand Banks of Newfoundland range as far south as 42 degrees north latitude, about 500 miles south of Newfoundland, causing a serious danger to shipping. (Antarctic icebergs are longer

—up to 60 miles! And, although their average northern boundary is 45 degrees south latitude, individual pieces of ice have been seen as far north as 26 degrees 30 minutes south—as far as mid-Australia.)

Different sensors can be used for making pertinent measurements of sea ice and ice-ocean interfaces. Thus, microwave radiometers are particularly useful in detecting the boundaries of pack ice because of the ice's inherently higher radiant temperature as compared with the surrounding water. Passive microwave radiometry measurements in the 1- to 2-cm band may be useful for working out the thickness of relatively thin sea ice; however, thick, old ice with varying salinity concentrations in its many layers is not susceptible to such measurements, because the rate at which heat radiates from it is not constant. For cloud-free reconnaissance over sea-ice areas, infrared sensors and color photography are indicated. Infrared imagery, for example, may reveal changes in ice thickness by measuring surface temperatures, resulting in profiles of relative thickness over the ground track. Because so many things influence the measured surface temperature (including the ambient air temperature, the nature and extent of cloud cover, the extent of snow cover, local humidity, and Sun angle), the technique cannot provide accurate values of ice thickness.

Certain features of ice concentrations and individual ice masses such as icebergs can be disclosed by active microwave radar systems, which are especially useful when there is extensive cloud cover and during the long polar nights. Not only can radar make known where ice is located, but it also can yield data on the type of ice and, from an analysis of the strength of the radar return, its thickness.

Such are some of the actual and potential advantages of the application of artificial satellites to oceanography. Long gone are the days when one could speak of the oceans as a "gray and melancholy waste." Today, scientists speculate that the very future of man as a species may depend on how efficiently and how wisely he utilizes—and cares for—the vast open waters that cover nearly three-quarters of the surface of our planet. Just how he will put their bounties to the expanded service of mankind is still not clear in all its details, but it is certain that the instrumented satellite—and eventually the manned space station or modular elements associated with it—will play a leading, if not decisive, role.

6
THE LAND

TWENTIETH-CENTURY MAN IS A PRODIGIOUS consumer of energy and natural resources, largely developed on or derived from the land surfaces of the Earth. The United States, Western Europe, the Soviet Union, and Japan all maintain standards of living closely related to their production of goods and services, and less-developed parts of the world struggle to industrialize so as to improve the lot of their peoples. Rising human ambitions, coupled with a staggering growth in sheer numbers, already are straining the basic resources of our planet, and within the relatively near future selective mineral scarcities and even exhaustions may occur. Closely tied to mushrooming consumer expectations is the specter of an environment contaminated by the accumulation of filth that high living standards create.

By the year 2000, if the world is lucky, clean solar energy may supplant that derived from fossil fuels (coal, petroleum), radioactive elements, and water—all of which disturb the environment while being converted into useful energy. In the meantime, these conventional energy sources will continue to be used to manufacture the materials, comforts, and services on which the economic well-being of our civilization rests. Other than energy, man requires minerals from which many of his myriad goods are processed. Despite great efforts on the part of mining and petroleum enterprises to locate new resources, discoveries by traditional means are not being made at a rate capable of meeting anticipated demands by the year 2000.

Another resource, clean and abundant water for drinking, recreation, irrigation, and industry, also is becoming scarce. And agricultural and forest resources may not long be able to match the voracious appetites of the world's billions. Starvation still occurs despite the "green revolution" made possible by the development of improved wheat and rice strains; in Yemen, for example, some 2 million people were threatened with famine as the 1970s decade got underway. As for so-called cultural resources, the works of man on the face of the Earth (his highways, railway lines, airports, cities and towns, factories, and the like), there is doubt that they can continue to serve humanity to the extent that they have in the past. Cities are becoming polluted, crime-ridden centers that breed discontent, despair, and fear. Old roads are clogged and new roads eat up invaluable agricultural and recreational land at an alarming rate. Airports are often so crowded that long takeoff, landing, and baggage delays are common. And the problems of getting between airports and the urban centers they serve increase daily. Because of uncontrolled suburban, highway, and other building expansion, land-use maps are all too often outdated just at the time when man needs a convenient and accurate source of information concerning how his lands and their resources are being employed, on which he can plan immediate, near-term, and long-term agricultural, forestry, water, geological, and other activities.

Fortunately at this critical juncture in the chronicle of civilization, a remarkable new tool has become available: the land resources satellite. Historians of the future may compare this development with that of fire or the wheel, so revolutionary and so basic may become its ultimate impact on human society.

Land resources satellites, manned or unmanned, can undertake an amazing variety of studies, ranging from remote explorations of economic mineral deposits and wide-area studies of geologic structure to observations and measurements of irrigation water; inventories of lake, river, and stream waters; measurements of lake and estuarine water pollution, thermal and ice conditions in inland waters, groundwater discharge, and snow density and melting rates; topographic mapping; forest and range fire detection; general resources utilization; and detailed studies of the growth and vigor of crops and the optimum time for harvest. Volcanic activity can be detected from orbit and earthquake surveillance—and perhaps even prediction—undertaken by the continuous monitoring of instruments installed on the surface that measure ground tilt, strain, seismic activity, and possibly fluctuations in the local magnetic field. The extent of floods can be determined almost instantaneously from orbit, leading to more efficient and rapid rescue and supply activities on the ground below. And, by observing the rise of waters upstream, the probable future course and degree of flooding can be estimated.

From the viewpoint of area, the study of the land is less ambitious an undertaking than the study of the much larger ocean surface of the planet. Yet it involves masses from the size of Asia, with more than 17 million square miles, to the tiniest, uninhabited Pacific atoll. It is also the study of a mountain like Everest more than 29,000 feet above sea level and of the Dead Sea 1,302 feet below sea level, of such vast megalopolises as the Boston–New York–Philadelphia–Washington corridor in the United States, of deserts such as the 3.5-million-square-mile Sahara, of lakes such as Superior (31,820 square miles in area), and of rivers—for example, the 4,157-mile-long Nile. In short, it comprises the study of everything on the surface of our planet not covered by oceans and seas.

The techniques for studying and surveying the land from orbiting satellites are the same as those used for the oceans, and include:

1. Placement of automatic sensors on the ground or in balloons and airplanes above the ground. As a satellite passes overhead, the information gathered by the sensors is read out via a telemetry link. The satellite can either store the data or relay them immediately to another satellite or to a ground station within its field of view.

2. Installation of automatic cameras aboard a satellite to take pictures of the land below at preselected intervals, or according to the occurrence of some event, or in response to command from the ground. The pictures are then stored until such time as the reentry capsule is jettisoned from the satellite, passes through the atmosphere, and is recovered either by aerial "snatching" or after landing in the ocean. Satellite payload recovery is now a routine, reliable operation.

3. Incorporation of remote sensors, including television cameras, in

the satellite. Upon command, or at programmed intervals, accumulated data are transmitted, or "dumped," to ground control stations as the satellite orbits overhead.

Land resources satellites use a variety of remote sensor techniques, depending upon the phenomenon to be measured. As with remote sensing of the vast ocean bodies, spectral-signature research applied to the land involves such factors as absorption, emission, and reflection of energy—all of which vary as functions of time, season, and climate. Moreover, many land phenomena can be examined in more than one spectral region, leading to composite images useful to the geologist. As with the oceans, thermal sensing in the infrared region is applicable to certain land phenomena, especially such short-lived and relatively uncommon events as forest fires and volcanic eruptions. Finally, the advantages of microwave sensing can also be used in land reconnaissance, making all-weather and nighttime imaging practical.

Five important land sciences benefit from resources satellites: (1) geography, including cartography; (2) geology, including geologic mapping; (3) hydrology; (4) agriculture; and (5) forestry. They overlap to an extent; for example, the problem of soil erosion and conservation is of interest and concern not only to the agricultural field but also to the others. The extent of soils and the broad questions of soil use are the domain of the geographer, including the economic geographer. The origin, development, classification, as well as the physical nature and properties, of soils are all inherent to the field of geology. Many of these elements, erosion and conservation in particular, deeply involve the agronomist or the forester. But the hydrologist must also deal with the effects of erosion: silting of harbors, filling in of navigable rivers, disturbances to watersheds, to mention but a few.

GEOLOGY AND GEOGRAPHY FROM ORBIT

Partly because of the critical need for expanding reserves of natural resources, partly because of the high cost of conventional exploration techniques, partly due to general economic pressure, and partly in response to intellectual stimuli (the interest of the theoretical geologist), the pace of geological research has accelerated markedly in the past decade or so. And so have investigations into the feasibility of applying Earth-orbital sensing and observational techniques to the geosciences, with the hope of reducing the costs of, and increasing the yields from, searching for such resources as mineral fuels, nonmetallic deposits, and metallic ores.

The advent first of instrumented unmanned satellites and later of manned orbiting spacecraft is particularly important since it is no longer common to discover exposed deposits of economically valuable ores. Rather, exploration theory must be called upon to provide the clues and guidelines which can lead to the recognition, and later assessment, of structural and lithologic conditions that favor the occurrence of useful but hidden extractable deposits. Once this has been accomplished, conventional geophysical, surface geological, geochemical, and laboratory methods can be brought to bear.

The study of geology can be approached from many angles and from many points of view, from the macroscopic to the microscopic. Obviously, laboratory analyses of rock samples cannot be conducted from orbit. Possibly less obvious is the fact that much geophysical surveying by gravimetric, magnetometric, seismic, and electrical methods will continue to be undertaken effectively from the ground or the air—even though some instruments are beneficially operated from orbit. But broad geological and geophysical surveying—the study of features of major rock masses, of tilted and folded strata, of fractures and fracture structures, and of topographic expressions as they affect geology (for example, valley patterns, which depend on the distribution of bedrock, the arrangement of various surfaces like faults and joints, and the attitude of stratiform rocks)—all of this is amenable to satellite sensing.

More specifically, geologists look to geosatellites for photographic outputs (either by recoverable film capsules or by television) from which they can prepare regional and continental geologic photomaps, gross lineation or lineament maps (large-scale linear features of regional dimensions), structural geologic maps, and synoptic geologic maps. Once these have been completed and the necessary interpretations of them made, promising areas can be selected for conventional field geological exploration. Also, following geochemical analysis, local subsurface structures can be studied by cable-tool drilling,

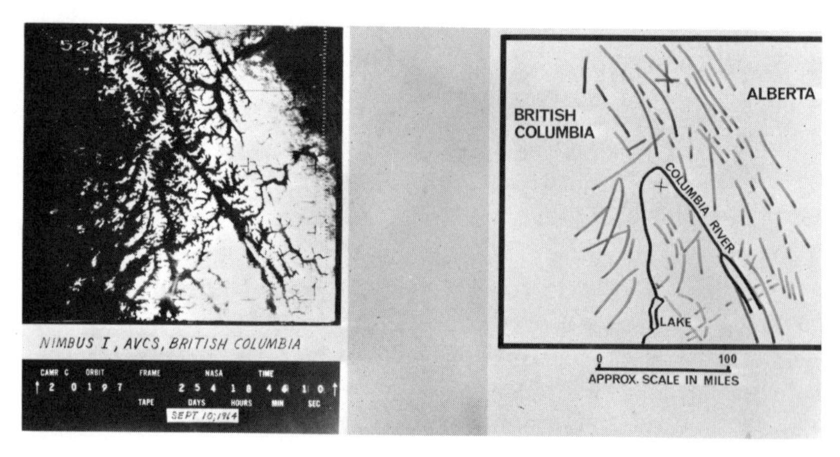

Light snow cover assists geologic observations from orbit by enhancing stream drainage patterns. Sketch map at right elucidates linear geological features in relation to the drainage system shown in a Nimbus 1 photograph of British Columbia and Alberta. (NASA AND U.S. GEOLOGICAL SURVEY)

rotary drilling, diamond drilling, side-wall sampling, and a variety of other techniques.

The importance of geologic mapping to field and economic geologists is overriding, yet traditionally they have had to spend months or years conducting their investigations from the ground—and, relatively recently, with aerial photographic support—patiently waiting for the geologic picture to emerge. Making and discarding hypotheses as he works, the geologist moves toward increasingly accurate interpretations based on what he seeks and what he surmises, always devising tests to check them out. Obviously, experience and judgment are of inestimable importance in this process.

The geologic map is the basis on which the search for economic minerals begins. Frequently, government topographic and geologic survey maps serve as a foundation on which ore and mineralization data can be plotted. However, many regions of the world are inadequately mapped geologically, forcing exploration groups to make their own maps. Insofar as possible, work on the ground is supplemented by aerial photographs. The latter are extremely useful in locating features not immediately or easily observable on the surface, as, for example, faulting—often revealed from above by variations in soil color and vegetation. Other topographic forms—fractures, tilted and folded structures, and related geologic expressions—that may be delineated (and even discovered) from above are revealed by valleys, basins, ridges, reefs, hoists, grabens, batholiths, scarps, cliffs, beaches, terraces, cuestas, and hogbacks.

Once the mapping of a new area is completed, mining geologists will note: (1) those tracts that either are known to contain ore bodies or give evidence of conditions favorable to economic deposits; (2) those where the structure is but moderately promising; (3) those where favorable conditions have not been found; and (4) those where such conditions are fairly certain to be absent. Their annotated maps, accompanied by detailed geologic reports (and often the results of laboratory analyses), provide the framework within which valuable resources are uncovered. Among the resources most amenable to regional surveys are iron, copper, and gold ores; oil and gas; and such nonmetallic deposits as limestone, gravel, and sand.

The principal advantages of the satellite overview to the exploration geologist are the synoptic small-scale coverage he is able to obtain; the benefit from uniform lighting conditions over large areas; and the reduction or elimination of interference to his force field measurements. The observation and sensing of phenomena over a broad area at a given time allow him to interpret the resulting data without having to make local adjustments for climatic factors. In the past, in order to prepare a mosaic of a large region of geological interest, it was necessary to compensate for seasonal differences which give rise to changing vegetation, moisture content of soil, and snow cover. Because the synoptic overview was previously unavailable, studies had to be based on an amalgam of large-scale observations made on the ground or from the air, spaced over extensive periods of time. The overview now makes it possible to secure small-scale coverage of vast geologic regions of the globe (the entire Andes chain, for example, or a complete country like India) and to recognize broad structural anomalies—capabilities of inestimable value in mapping. Errors arising in the preparation of mosaics, from reduction of large-scale photos or maps and from syntheses of numerous and intermittent ground and aerial observations, are reduced.

Directly correlated with the overview is the advantage of uniform illumination of extensive geographic features. Formerly, daily and seasonal lighting variations reduced the accuracy inherent in a broad

Clear photographs of isolated geographical areas, such as this one of the Baja California peninsula, taken by astronaut L. Gordon Cooper during the Gemini 5 mission in 1965, help cartographers make and update maps all over the world. (NASA)

"single-look" observation, and led to a degraded understanding of those geologic features that are enhanced by shadows and uniform illumination when viewed from orbital altitudes.

As far as "noise" (interferences or disturbances) is concerned, by surveying the Earth from above, anomalies and variations affecting surfaceborne instruments are either lessened or eliminated. Typically noise is the result of localized magnetic and gravity-field variations. Related to this is the fact that irregular ground and aircraft observations are not likely to provide the most accurate data possible on features affected by short-term variables.

Some of the principal techniques used in geologic mapping of major structures by land resources satellites are regional color photography (as accomplished in Mercury, Gemini, and Apollo flights), infrared surveying, and radar imaging. Such work can lead to the discovery of important economic resources, including petroleum, natural gas, and metallic and nonmetallic minerals. Full advantage is taken of "shadowing" or "shadow enhancement," wherein low-angle solar illumination highlights the vertical dimensions of geologic features and surface patterns.

Geologic mapping from orbit, besides revealing new sources of minerals, may yield information for planning major construction efforts, for example, dams and bridges, which are influenced by geologic factors and which in turn influence large geographic provinces. Orbital surveillance can reveal such potential dangers as unstable water-saturated soils, fault zones, and landslide areas.

Land resources satellites are ideal platforms from which to monitor thermal activity, including volcanoes, on the surface below. Continuous surveillance in the visible and infrared regions of the spectrum may reveal thermal anomalies, possibly associated with crustal movements, that may anticipate volcanic eruptions, earthquakes, or landslides. New insights may result as to the overall internal thermal activity of the Earth. Moreover, orbital reconnaissance could well lead to the discovery of new geothermal power sources. And, by repeated monitoring of snow thaws, undiscovered mineral deposits may be inferred because they occasionally produce enough heat in the course of oxidization to affect melting patterns. Data from ground-installed radiometers and tiltmeters could also be relayed through a land resources satellite, contributing to a large-scale thermal monitoring program.

Similarly, seismic instruments and strain gauges could be placed in myriad locations all over the world to monitor seismic activity. Their outputs would be relayed by radio link to the satellite(s) and thence to ground interpretation centers. Meanwhile from orbit, photographic and imaging devices could provide continuous surveillance of major tectonic regions to improve our understanding of earthquakes. The U.S. Geological Survey's National Center for Earthquake Research believes that at least short-range prediction (hours or up to a few days) is possible. The late spring 1970 quake that devastated a 600-mile belt on the Pacific coast of Peru killed an estimated 50,000 to 60,000 persons, injured at least 100,000, and left some 800,000 homeless. At least $500 million in property and other damage was caused —a potent reminder of the need for research in the forecasting and prevention of natural disasters.

Magnetic and gravimetric activity can be monitored from orbit, helping the exploration geophysicist relate force-field data to rock and terrain features. Magnetometers carried in satellites can detect broad magnetic anomalies, which may lead to the discovery of buried ore deposits and petroleum reservoirs. Gravity-field information is obtained by measuring the perturbations of the satellite as it circles the Earth, though devices designed to make direct measurements of the gravimetric field—the rate of change in the intensity of gravity per horizontal unit of measurement—may prove useful. Fortunately, from orbital heights, minor near-surface anomalies are minimal; they therefore have little effect on the determination of major magnetic and gravimetric anomalies deep in the Earth. There is strong evidence of the correlation between such anomalies and the presence of important mineral districts.

Of prime importance to the success of a geosatellite program is the full understanding of the physics of measurable geologic phenomena, on the one hand, and the methods and procedures of remote sensing, on the other. For example, it should be possible to make comparisons between images recorded simultaneously in several spectral bands so that ground phenomena can readily be revealed. Gradually, information will be accumulated on rock, soil, and mineral "signatures" at various wavelengths and under many environmental conditions, leading to new insights into the origins and relationships of mineral and petroleum provinces.

Recognizing that observational data are dependent on frequency and wavelength, and that electromagnetic radiation is attenuated—to varying extents—by the media through which it travels, one must determine, over an extended period of time, just what individual remote sensing measurements involve. What, for example, is the depth of penetration of a given electromagnetic wave? (Radar penetration depths run from 1 to 10 cm.) To assure the accuracy of the coupling between remote sensors in orbit and rocks, minerals, and soils on or below the ground, the same sort of "ground truth" program required for the remote sensings of the oceans must be applied to geology.

Although land resources satellite technology is still in the developmental stage, it is possible to provide preliminary (and doubtless conservative) estimates of the general order of financial savings that could result from an operational system. The United States consumes natural resources at a prodigious rate, and will probably double its present annual consumption within the next two decades. Even though it is richly endowed with its own natural resources, like other highly industrialized countries its import requirements become greater with every passing year. This implies an interest in increasing not only proven internal reserves but also those of less industrialized countries on which it is dependent for imports. The geosatellite offers an important new—and economical—tool to meet these objectives.

It is estimated that the United States and Canada together spend over $600 million each year on geologic, geophysical, and mapping operations directly related to the search for oil and mineral deposits. To this is added some $65 million spent on regional geologic mapping and general geophysical studies. Supposing—and only supposing—that the data collected from exploration activities were useful for only a decade; then even a 1 per cent improvement in efficiency from the use of geosatellites would make the system most attractive—yielding well over $60 million. Yet the U.S. Geological Survey has estimated that such a system would produce an improvement in exploration efficiency of 7 per cent. Whatever the accuracy may be of this particular estimate, the use of remote geologic sensors almost certainly will stimulate the discovery of new and valuable oil and mineral reserves, whose value may prove comparable to the discoveries made as a result of breakthroughs in exploration techniques in the past. And the new method of exploration will cost far less than existing ones since much of the cost of developing the necessary launch vehicles and spacecraft already has been paid in other aerospace programs.

Although the ultimate consumer is the direct beneficiary of land resources satellites, many of the geologic data furnished by them are first utilized by such government organizations as the U.S. Geological Survey, the Bureau of Reclamation, the Bureau of Mines, the Army Corps of Engineers, and the Environmental Science Services Administration; by various state agencies; by mining, petroleum, and engineering-construction companies; by universities and research institutes; and by individual consultants. Similar organizations in other countries would benefit from international programs. When leading scientists and engineers are warning of impending widespread shortages of energy and of many vital natural resources, the importance of the land resources satellite is easy to comprehend.

MONITORING HUMAN ACTIVITIES FROM ORBIT

The actual and potential uses of satellites monitoring human activity are many and varied, as seen in the typical listing given here. Almost all require repetitive coverage so that the dynamic interface between man and his cultural resources can be better understood. If the many, extremely serious problems facing mankind—the destruction of his environment, the exhaustion of his natural resources, an uncontrolled population—are to be resolved effectively, continuously updated, synoptic information will become ever more important. No rational solutions to problems such as these can be found without complete and timely data available to planners, not simply on a regional or national but on a worldwide basis. What man does in one part of the globe can have important, and often dangerous, consequences everywhere. The soot from the industrial revolution is entombed in the Antarctic icecap, and penguins in that remote corner of our planet harbor pesticides in their internal organs.

The economic advantages of remote sensing and manned observation from orbit of human activities are too vast to be assessed at this early stage. But inevitably the day will come when man will find it difficult to realize that he ever tried to manage a planet without the advantage of the overview. Notwithstanding the complexity of placing dollars-and-cents values on the myriad benefits looming on the horizon, one can cite a few figures. To map the Earth's 58 million

TYPICAL CULTURAL APPLICATIONS OF
LAND RESOURCES SATELLITES

Determination of the actual population boundaries of urban centers, and boundaries of ancient cities

Rates and directions of change of these boundaries

Types of change, for example, deterioration of cores, urban renewal, suburban spread

Global movement of populations

Instantaneous status of surface and air transportation

General inventory of geographical and man-influenced resources and rates of change, for example, the removal of agricultural land from production due to the construction of a new shopping center

Control of short time-base phenomena, such as dust storms, forest fires, and burning off of open fields

Detection of air and water pollution and monitoring of pollution violators–a factory beside a river or lake, for instance

Changes in the energy budget caused by man's activities, especially around great urban centers

Disturbances by man of natural geomorphic forces, as soil erosion, silting of rivers, filling in of lakes

Observation of natural and man-made catastrophes

Aid in planning large-scale enterprises such as laying a pipeline (for example, the most feasible routes from a geographic point of view) or a river control project (for example, probable geographic, ecological, and cultural effects, both beneficial and detrimental)

---

square miles of land surface using aerial photographic techniques, an estimated cost of $174 million was given in 1966, compared with only $17 million if space systems were used. The figure does not include the cost of the carrier vehicle and the satellite, since it was assumed that the latter would have been orbited anyway for other purposes. (The addition of an experiment to a satellite launch mission has become a routine matter, so long as the experiment does not interfere with the prime purpose of the mission.) The indirect benefits that would accrue, merely from the availability of up-to-date maps in the United States, would result in a savings of at least $136 million

Two Apollo photographs were enlarged, rectified, mosaicked, and overlaid on a standard 1:250000 topographic map of the Phoenix area to produce this updated "space photomap," which contains far more information than the standard line map. (U.S. GEOLOGICAL SURVEY)

each year. If the program were applied worldwide, the annual benefit could exceed $10 billion.

Geographers estimate that even by 1970, less than half of the land surface of the planet Earth had been adequately mapped. Many areas either have not been mapped at all or have received only minimal attention from the cartographer. Often these are the very ones where important economic advantages for underdeveloped countries can be realized. And even where fairly detailed maps are available, they are usually obsolete; the nations involved may not have the economic resources to update them at regular intervals. Even some parts of the United States have not been photographed in 20 years, and many "modern" maps are a decade old. In highly developed countries the use of land can change strikingly in just a few years, making old maps nearly useless. One need only consider the advent of the interstate highway system to realize this.

Mapping from the ground, and even from aircraft, can be very time-consuming and costly, and usually involves masses of data that must be handled, stored, and interpreted. Exploding populations, the rising expectations of nearly every member of humanity, and dwindling known reserves of some natural resources combine to make the creation of modern, detailed maps one of the most important objectives of the age.

The advent of photography from airplanes to a great extent replaced the ground surveyor. Now aerial photography inevitably is yielding to spacecraft photography, as a new revolution in cartography is in the making. The limitations of the airplane are several and obvious. It is restricted by altitude, range, and speed constraints; it is relatively ineffective in inclement weather; and, for operations in remote areas, it requires specially constructed bases from which to operate. This last implies the need for handling crews, mechanics, other specialized personnel as well as the many facilities and equipments they require, and—inevitably—high costs. An excellent example of a remote air installation from which much mapping takes place is Williams Field near McMurdo Station in Antarctica.

Even when aerial mapping operations are financially feasible, altitude limitations prevent synoptic coverage. Moreover, the coverage that is possible is necessarily discontinuous and, especially in areas of difficult access, is widely spread across not only seasonal boundaries but time intervals that may approach decades. Today, to make a standard small-scale map of a country like the United States, about 200 stereopairs of photographs need to be taken, at about 30,000 feet altitude. Since some five picture control points are needed for each one of the stereopairs, approximately 1,000 photo control points are required for the map. To put together a new large-scale map of the United States, about 1 million photographs would be needed for the photo mosaic. Several years at least would be required to assemble the pictures.

Synoptic photography from orbit, in contrast, can yield the same results as the 200 stereopairs with only a single stereopair produced by a metric frame camera and three stereopairs by a high-resolution camera to provide the necessary map detail. Moreover, by employing a narrow-angle lens, photographs from orbit would not suffer distortion. Instead of 1 million pictures, some 400 would do the job of covering the United States, and they could be assembled in only a few weeks. In addition, complex rectification processes would not be needed.

Satellite photographs have provided the first comprehensive view of the geography and geology of Libya. Dominating this Gemini 11 photo is the circular Murzuch sand sea. In the foreground are the rugged Tassili n'Ajjer mountains; the Mediterranean Gulf of Sirte is visible at top. Algeria is to the left; Egypt, to the right. (NASA)

An outstanding, but by no means unusual, example of the application of synoptic terrain photography to the geology and geography of a poorly mapped region is found in Angelo Pesce's *Gemini Space Photographs of Libya and Tibesti*. According to the author, ". . . observations made by explorers in the early decades of the twentieth century are still incorporated into some recently published maps." Only when photographs from the Gemini 5, 7, 9, and 11 flights became available did some virtually unknown and unexplored geographic and geologic features finally come to light.

On the basis of the Gemini photographs, basins and uplifts were outlined, sedimentary and igneous outcrops delineated, and regional unconformities and lithological units traced. Many geomorphological and geographic features and relationships were observed and understood for the first time, and the topography recorded of such unmapped areas as southern Cyrenaica between the Calanscio "sand sea" and the Egyptian, Sudanese, and Chad borders. In the northern Tibesti foothills, Pesce noted many leveled-off circular features believed to be—for the most part—granite intrusions leveled by erosion. Although much information of theoretical interest was gained, the fact that the project was sponsored by the Petroleum Exploration Society of Libya underscores the commercial interest in satellite mapping.

HYDROLOGY FROM ORBIT

The value to humanity of efficient water distribution is immense—for water is the very lifeblood of civilization. Consequently, even small improvements in the management of irrigation systems, regional and local water distribution, and industrial use of water yield great economic returns. Floods are estimated to cost the United States, on the average, $300 million a year; individual years have witnessed losses in excess of $1 billion. Hydrologic satellites hold great promise of reducing these losses, by relaying data from thousands of individual reporting and measuring elements of flood warning systems to control points, from where warnings can be sent out to the potentially affected areas. And after flooding is under way, hydrologic satellite observations will be able to show the limits of flood expanse so that corrective actions can be taken, damage assessed, rescue operations coordinated, and plans made to improve warning networks to reduce the probability of future catastrophes.

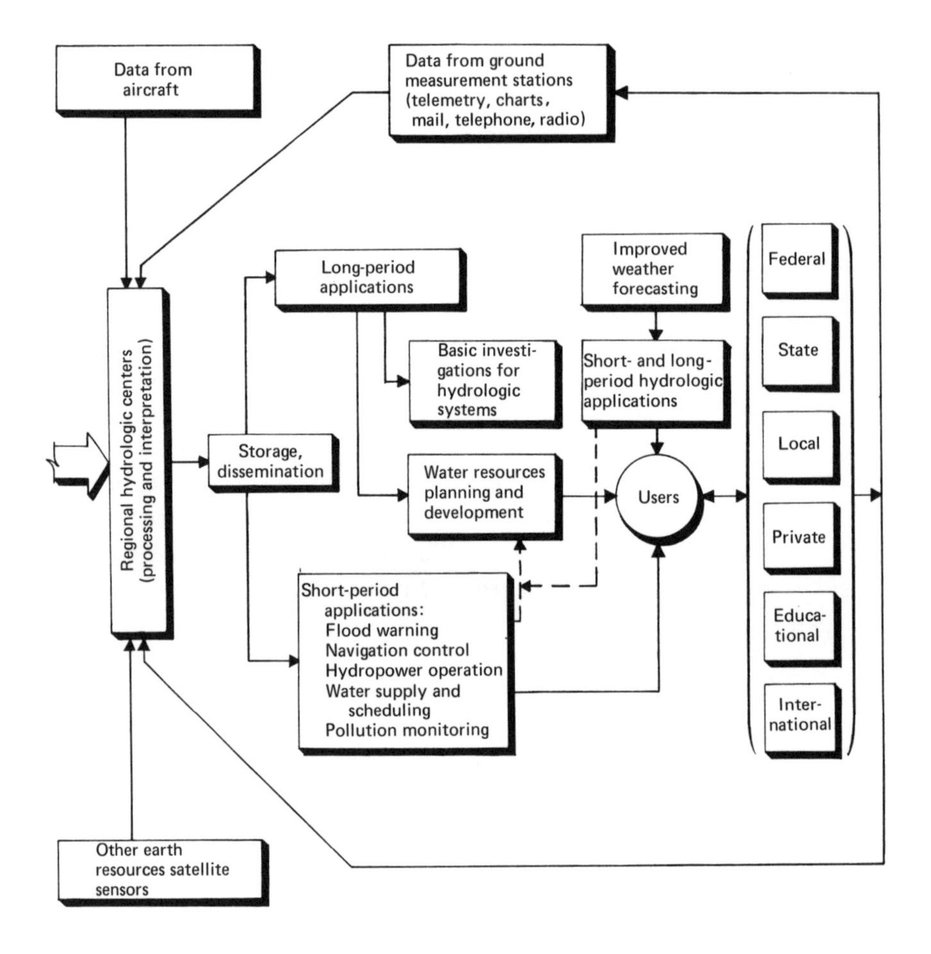

In the broadest sense, the satellite overview can help man to gain a deeper understanding of the entire hydrologic cycle and the varied influences that cause it to function as it does. With increased information available from direct orbital surveys and from ground-generated data relayed through satellites, man will better be able to ascertain the amount and quality of water stored in rivers, lakes, and reservoirs; as snow and ice; and in aquifers (water-bearing strata). With such knowledge it will be possible to predict more accurately the general availability of regional water and to provide it in good quality, at relatively low cost, and with minimal danger of local droughts or floods. Of special importance, in an era of broadening concern for

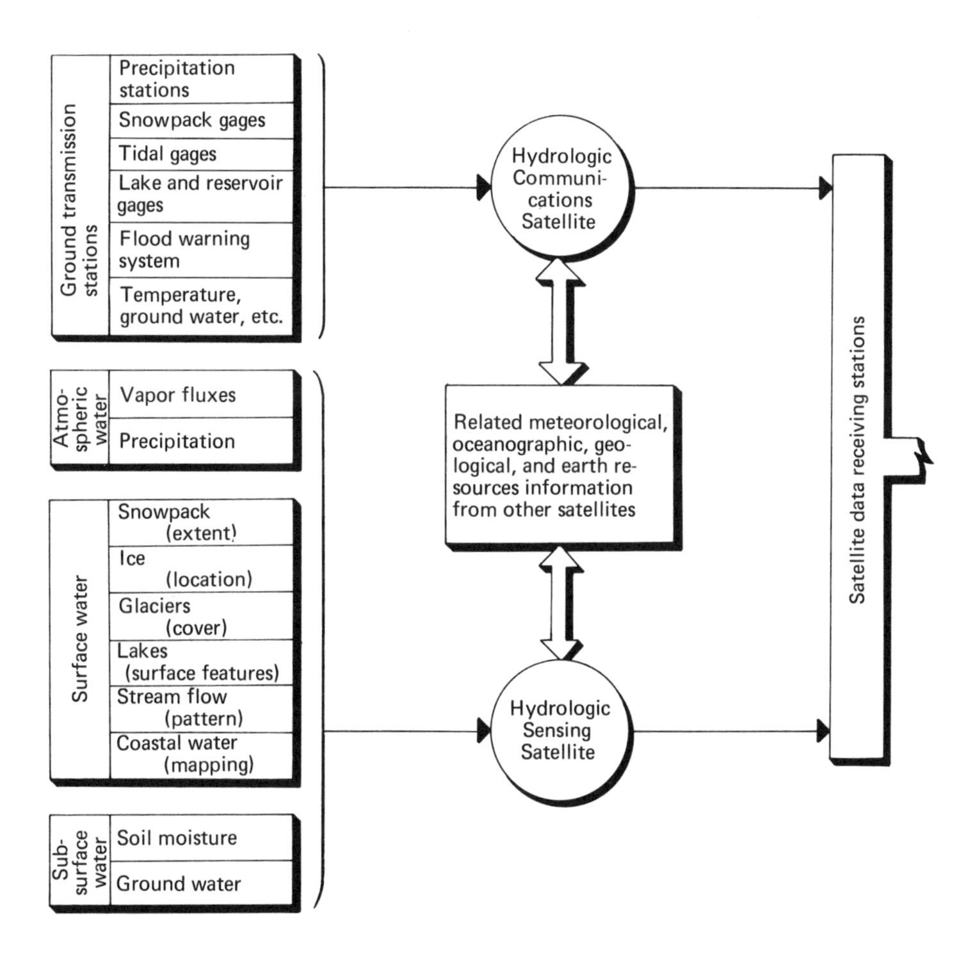

the environment, is that everything possible be done to minimize the ills of water contamination through the discharge of polluted effluents into inland water systems. The overview can not only spot and monitor pollution at its source but also follow its course downstream.

The goals of modern hydrologists are ambitious. They talk of establishing a global water-information service, along the lines of the World Weather Watch. They are also interested in applying to the fullest extent possible the integrated-systems approach to major hydrologic problems; and they would like to set up national and international training programs, in recognition of the fact that most water systems affect more than one country. To an increasing extent it has become evident that such goals can only be realized by the inauguration of a vigorous program of satellite hydrology—applying the fruits of space science and technology to the study and management of vital water resources in every corner of the world.

In principle, the same advantages accrue to hydrology from the satellite overview as to oceanography, geology, and other sciences. As in other capacities, satellites provide the hydrologist with repetitive, global data both by direct observation and by serving as data relays. Many of the actual and potential applications of hydrological satellites are summarized in the listing.

Typical Hydrological Applications of Land Resources Satellites

Preparing land use and land classification maps in terms of their application to the design and management of water resources systems

Monitoring worldwide, regional, and local water consumption; observing major diversions of water for industrial and other uses

Assessing, on a regular basis, current and predicted short-term water requirements (for example, for proper planning of water resources projects)

Planning for long-term uses of water as based on predicted requirements of expanding populations, rising industrialization, and urbanization

Conducting inventories of water stored in rivers, lakes, and snow and ice fields; monitoring water (including groundwater) levels

Studying the biochemical composition of river, lake, and other inland waters (for example, growth of algae)

Monitoring transport and deposition of sediment

Observing the extent of saltwater intrusion in fresh-water bodies

Scheduling flows of water to hydroelectric plants, for irrigational purposes, for flood control, and for urban and industrial consumption

Planning the design and construction of waterfront structures, based on knowledge of such factors as water circulation, wave action, probability of flooding, ice accumulation, and force of winds during ice breakup

Controlling beach, riverbank, and lakefront erosion (for instance, to protect waterfront structures)

Protecting inland-water fish and shellfish, waterfowl, and wildlife dependent (for breeding and other purposes) on such water reserves as swamps and marshes

Detecting pollution at its source, including illegal dumping of wastes; tracing its movement; observing dispersion of visible pollutants by recording turbidity and color changes; planning how to dispose of municipal and industrial waste effluents and heat

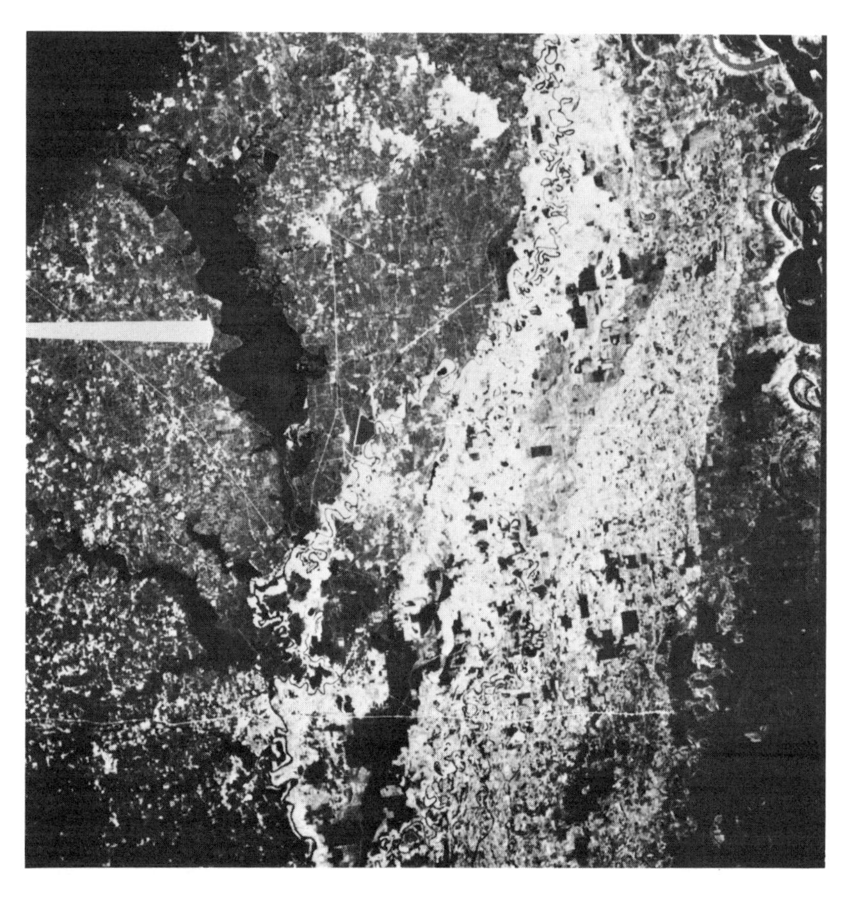

Louisiana and Arkansas, seen from Apollo 9 in March 1969. The generally lighter-toned alluvial area runs from top to bottom of the photograph. Clearly indicated in deep blue in the original color-infrared image is a flood along the Ouachita River (marked here by arrow). This was the first time an entire flood was seen in a single space photograph. (NASA AND U.S. GEOLOGICAL SURVEY)

Sensors employed for hydrology are similar to those used for other Earth resources: panchromatic and color cameras, infrared detectors, and radar equipment. The extent and intensity of rain in a given area can be established in theory by radar, but much experimental work needs to be accomplished through aircraft testing before the technique can be successfully attempted from orbit. Radar and microwave techniques may be useful to determine the depth and movement of inland waters, while microwave can yield data on soil moisture. By noting differentials in temperature, infrared detectors can record the buildup of water pollution in lakes, rivers, and streams. Of widest application is panchromatic and color photography, which shows clearly the patterns of flow in rivers and streams, incipient and actual flooding, the onslaught of droughts, and circulation in lakes. It also can delineate watershed drainage patterns; the amount of ice in reservoirs, lakes, and rivers, including ice jams; and the extent of snowpack. This type of photograph also is useful in determining the area, size, and movement of glaciers, and the intensity of boat traffic and logging operations on inland waterways.

As in the case of other Earth resources sensing devices, some of these hydrologic sensors are already proven; others, in the experimental stage, are being tested in aircraft; and still others have yet to be demonstrated.

It is estimated that at least 10,000 surface stations would be necessary in the United States simply to provide flood warnings by monitoring the stages of major river and lake systems. Such stations would sense a particular phenomenon and transmit the resulting data to an Earth-circling satellite for real-time or delayed readout to hydrologic control centers. From there, the pertinent information would be relayed to the area potentially affected. Other remote stations would be instrumented to record precipitation, snowpack, tidal movements in coastal areas, and pollution. Depending on the urgency of the information gathered, transmission would take place on command, according to preset intervals, or at intervals whose frequency would increase or decrease after a preplanned time or after some event had occurred—for example, a rise in pollution concentration above a preselected value. In already developed countries, the type of data would be similar to, but more extensive than, that now collected; in other words, coverage would be not only much wider but more dense and

Infrared satellite photograph of the Cleveland area clearly defines pollution of Lake Erie by sediment from the Cuyahoga River. (U.S. GEOLOGICAL SURVEY)

more frequent. In developing nations, hydrologic satellites could revolutionize the knowledge, planning, and utilization of water resources.

The mapping of snow and ice cover offers one of the greatest benefits emanating from a hydrologic satellite system. At any one time, snow covers from 30 to 50 per cent of the Earth's land area, glaciers about 10 per cent, and pack ice much of the inland waters, particularly in the northern hemisphere. The cost of obtaining the information needed by conventional aerial and surface means is great; indeed, it is often impossible or extremely difficult to obtain it at all. Yet such data are urgently required for hydrologists to plan and operate irrigation projects, hydroelectric plants, regional and local water supplies, and flood warning systems.

Because of the sparsity of reporting stations; inadequacies in communication with remote stations in inclement weather; laborious, often hazardous, and infrequent ground surveys in mountainous and other areas of difficult access; and relatively rare and often costly aerial overflights, synoptic, continuously upgraded data on snow and ice cover are seriously lacking, even in developed nations. Yet it is calculated that if such data were available only in the western United States, the annual savings to water users would run from $10 to $100 million a year. It is estimated that many millions of dollars are lost due to damage from ice jams, and that many millions more are wasted by disruptions of navigation caused by unpredicted ice buildups. Avalanches cause much loss of life and property, while faulty programming of water resources owing to limited knowledge of ice and snow conditions certainly results in further losses at the national, state, and local levels.

AGRICULTURE AND FORESTRY FROM ORBIT

With populations rising at a terrifying rate and with social instability growing, partly as a result of overpopulation and partly because too many peoples barely subsist on limited food supplies (the United Nations estimates that nearly 70 per cent of the world's inhabitants suffer from either persistent hunger or malnutrition), it is evident that the world's food and fiber resources must be developed to the maximum. Since the availability of suitable land is essentially invariable, failure to apply the latest advances of science, technology, and management to our agricultural and forest resources is, almost literally, to court disaster.

This is not to suggest that our planet can continue to feed and clothe an ever-greater population, nor that the indiscriminate use of natural resources by the several generations now inhabiting the Earth can be condoned. Morally, man owes as much to the future as to the present, though his past actions hardly indicate any great concern for posterity. But as he attempts to solve population, environmental, and other pressing problems, he must also work for a stable society through which such momentous objectives can be carried out. To a great extent, a stable society means a well-nourished society, whose people have a reasonable expectation of living a life free from the specter of starvation or bare subsistence.

Agriculture and forestry, like other life-supporting activities, can benefit from satellite observations—not only the photographic coverage possible from orbit but also remote sensing of spectral signatures of individual crops. Adequate maps of food and fiber resources are lacking in many parts of the world, and much of what information is available is derived from black-and-white rather than color photographs. Moreover, all too many maps—and the photos on which they are, in part, based—are obsolete or of limited use due to the paucity of information contained or the format in which it is presented. Yet the overview is essential for the efficient and practical planning of agricultural developments—the conducting of inventories and the study of specific methods to improve land use, crop yields,

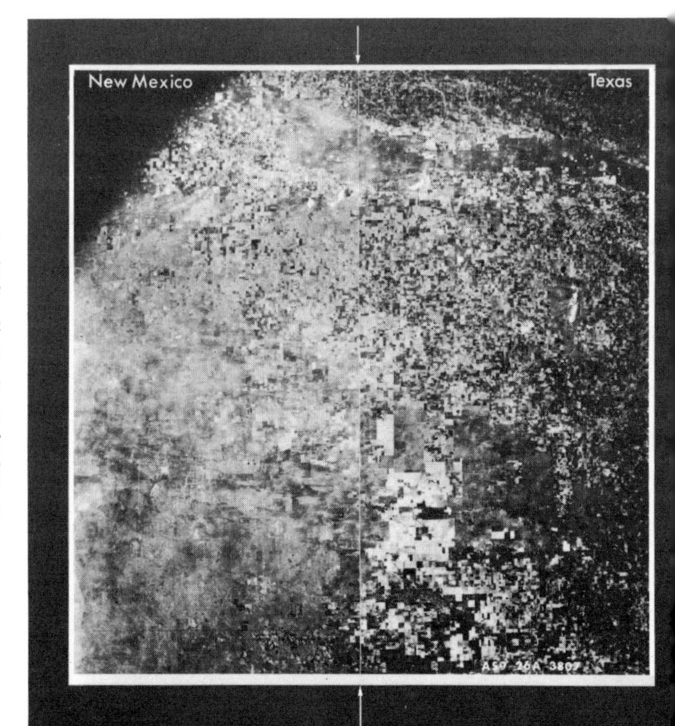

Contrasts in land use that result from differing state laws governing the extraction of groundwater for irrigation are apparent from an Apollo 9 view of the Texas–New Mexico border, even in this black-and-white version of the original color-infrared photograph. (NASA AND U.S. GEOLOGICAL SURVEY)

and irrigation programs. Likewise it is essential for the assessment of such factors as rate of crop growth, degree of insect and disease damage, and accuracy of soil classifications.

In order to determine individual spectral signatures of crops and trees, research must continue not only on how they reflect, emit, and radiate at various wavelengths but on the reflectance, emissivity, and radiance spectra of the soils in which they grow. Moreover, signatures for each and every type of growth must be related or considered in terms of the season; time of day; illumination; weather conditions between sensor and surface; quality of soil nutrients; local moisture conditions; disease, insect, and other stresses; and growth vigor. As in other Earth resources disciplines, ground truth must be carefully established before reliance is placed on satellite-sensed phenomena.

For general use requirements, three spectral bands appear appropriate: (1) the blue-green part of the spectrum, to permit haze penetration, shallow-water penetration for mapping underwater growths, and land-form mapping; (2) the red and near-infrared parts, for determining vigor of vegetation and moisture distribution; and (3) an intermediate band, to provide for crop recognition.

By examining visual images in combination with electronic images produced by sensors using several frequencies, tone and texture differences become evident, and many types of plantlife can readily be identified. Multispectral scanning distinguishes the diseased plant from the vigorous one because of differences in reflected or emitted radiation. Some infrared sensors can help evaluate crop response to different fertilizers used in varying quantities, as well as to pesticides, fungicides, and insecticides.

Although the application of satellite technology to the assessment and development of land resources is still in a preliminary stage, the expectations for even the short-term future are very bright indeed. Photographs taken from unmanned and manned satellites have been put to good use, and specific experiments already successfully carried out point to exciting things to come. In an age increasingly concerned with ecology and the environment, the Nimbus 3 experiment with a wild elk in the Grand Teton and Yellowstone National Park areas provides a timely and significant example. From an altitude of 700 miles, a 500-pound elk fitted with an electronic collar was successfully contacted and tracked to show migratory behavior, to gain information to body temperature, and to develop techniques that will make it possible to understand more fully the ecological stresses and constraints on wild animals in general and on endangered species in particular.

The gathering, compiling, processing, and interpreting of statistical information is stressed by agriculturists and foresters, who manipulate it in many ways so that they eventually can display it on maps indicating such factors as geographic location of crop and forest acreage, patterns of plant and tree growth, and the nature of soils. Small-scale photographs prove particularly useful in constructing topographic and regional soil maps, while large-scale photographs yield data on crop and tree inventories and the general capacity of land for agricultural purposes. (Most of the world's food and fibers comes from but 7 per cent of the land surface, with some 23 per cent devoted to pasture and range; the remaining 70 per cent is essentially nonproductive.) In the broadest view, orbital observations may help to make it possible to determine the optimum global population distribution and to help individual nations assess their own population control programs.

Without geosatellites, up-to-date, accurate inventories of world agricultural, range, and forest resources cannot be maintained, as traditional ground and aerial information-gathering methods are too slow, irregular, and imprecise. But there are many other services that such satellites can perform, as the listing demonstrates.

Typical Agricultural and Forestry Applications of Land Resources Satellites

---

Preparing land use and soil classification maps (classifying soils by color shade and organic matter content from orbit has been shown to correlate more than 90 per cent of the time with tedious on-site measurements)

Assembling worldwide, regional, and local inventories of crops, livestock, and fibers

Planning fertilization and pest and disease control of agricultural lands; verifying response of crops to use of fertilizers and pesticides

Detecting incipient and full-scale blights and insect infestations, to give farmers time to take preventive measures

Studying effects of urbanization—highways, airports, and other construction—on agricultural and forest resources

Detecting land recently abandoned, and intentional or unintentional land burning

Making improvements in range inventory and management, leading to increase in carrying capacity (including water resources management in areas of widely changing climatic conditions)

Warning of range disasters (fires, floods, erosion, wind-blown sand and dust buildup, predators, diseases, insect infestations)

Determining effect of air pollution on growth rates of trees

Detecting forest fires; assessment of destruction

Assessing forest yields, optimum cutting rates, reforestation programs

Observing wildlife preserves, habitats, and rookeries; taking wildlife censuses

Surveying recreational areas, to report on encroachments by builders, unauthorized removal of timber, health and vigor of vegetation, out-of-season hunting, and general administration of public lands

Predicting optimum times for cultivation and planting

Formulating a closer relationship between supply and demand, by determining the most efficient means of distribution, storage, and transportation (as based on expected crop, fiber, and livestock yields) and the most favorable selling markets—local, regional, and international

Planning movement of manpower, equipment, and services to harvest crop, forest, and range yields

---

The benefits derived from satellites tend to increase as the cost of gathering the raw data decreases and the uses multiply. In most cases, however, even approximate dollar values are hard to come by, partly because of the sheer volume of agricultural enterprise in the world. Of course, one can cite the cost of an individual forest fire; for example, the Coyote fire of 1964 near Santa Barbara, California, burned some 67,000 acres of property and forest land at a cost of more than $20 million, including $2.5 million just to extinguish it. Fortunately, airborne infrared mapping techniques were able to spot the main fire center through the smoke, leading to effective suppression. Forestry and firefighting experts estimate that the losses would have been $9 million greater if the infrared system had not been pressed into service at a critical time. Some forest fires involve heavy loss of life, as the October 1918 fire around Cloquet—and 24 other towns—in Minnesota, when 559 persons perished.

Forest fires not only destroy a salable inventory but kill wildlife, remove vast areas from recreation, cause the soil to deteriorate, invite the invasion of insects and diseases, and increase the runoff of surface waters, creating erosion. The immediate detection from orbit of a fire and the transmission of data on actual and predicted weather conditions (including surface and high-altitude winds) from meteorological satellites can help firefighters plan their attack.

Even more serious a danger to forests than fire are insects and diseases—about 3½ times greater in the United States, according to the U.S. Forest Service. Bark beetles, spruce budworms, and the chestnut blight have taken a huge toll from the nearly 509 million acres of commercial forest land in the United States. Synoptic, repetitive orbital surveys hold promise of detecting the advent of insects and diseases at an early stage of development. Similarly, such surveys could detect nutrient deficiencies and other data important to forest management.

Conservative estimates of the potential annual value of geosatellites to forestry are in the tens of millions of dollars, for the United States alone. In underdeveloped, forest-rich countries, the potential is far greater. An example of such savings, to which no precise monetary value was attached by the U.S. Forest Service, comes from experience, involving the use of Apollo 9 photographs showing a 5-million-acre Mississippi Valley area running across the states of Mississippi, Arkansas, and Louisiana. The interpretation of these photos brought about a reduction in the expected error of estimated timber volume of from 13 to 31 per cent. Savings resulted from elimination of normal flight surveillance, of complex interpretation of low-altitude photography, of field crews who would ordinarily have had to measure sample tree volumes, and of general maintenance in support of ground and aerial operations.

The making of soil classification maps for planning and management purposes is a costly enterprise, yet one that produces in the United States alone an annual benefit per acre of $5 for irrigated farmland, $1 for nonirrigated farmland, and $0.15 for rangelands and forests. This amounts to more than $860 million per year. Expanded to a worldwide scale, overall benefits would be valued at greater than

$7 billion a year. It appears that the only feasible way of making such global maps is by orbital surveying; other methods would simply be too costly and protracted to be practical.

Satellite monitoring of soil moisture presents a specific example of the value of soil information. In the case of a single crop—cotton, with an average of nearly 14 million bales produced in the United States each year during the 1960s—about 40 per cent of the yield came from irrigated lands. The Department of Agriculture estimates that if a 10 per cent improvement in information on soil moisture conditions of these lands were available from orbital surveys, the dates for irrigation could be optimally selected and the total quantities of water to be injected into the cotton-growing area established. It is estimated that this would result in a yearly savings of $100 million.

If observation and sensing techniques can be successfully applied to weed infestation of croplands (losses from which approach $4 billion a year), savings of at least 10 per cent ($400 million) seem possible. The detection of insects and diseases in cropland areas offers great benefits also; diseases alone cost nearly $4 billion annually in the United States.

The U.S. Department of Agriculture estimated that there were 109,661,000 cattle in the United States in 1969, of which more than 35 million were on ranges. If remote sensing were to be applied to range management—helping to assess areas of overgrazing, insufficient nutrient, accumulations of weed and brush, and the like—it is calculated that the land could carry approximately 10 per cent more animals. Modern range management permits the stocking of rangeland to about 85 per cent of its carrying capacity; but with improved understanding of range conditions and continuously updated information available from orbit, stocking could rise to 95 per cent. If 3.5 million more calves were produced annually, the benefits would amount to at least $350 million a year in the United States.

Similar surveys could be made of acreage devoted to crops, in terms of maximum suitability; maximum yield potential; optimum growth patterns for local, regional, and foreign markets; and storage and transport costs. The $40-million annual expenditure of the Crop Reporting Board, a branch of the Statistical Reporting Service of the Department of Agriculture, could be lowered by the receipt of such

Wide-area satellite photographs give geologists a new view of the Earth. The well-developed pattern of fracture lines in this Gemini 4 photograph of southwestern Saudi Arabia and Yemen suggests the presence of underlying domal structures that might contain oil. (NASA)

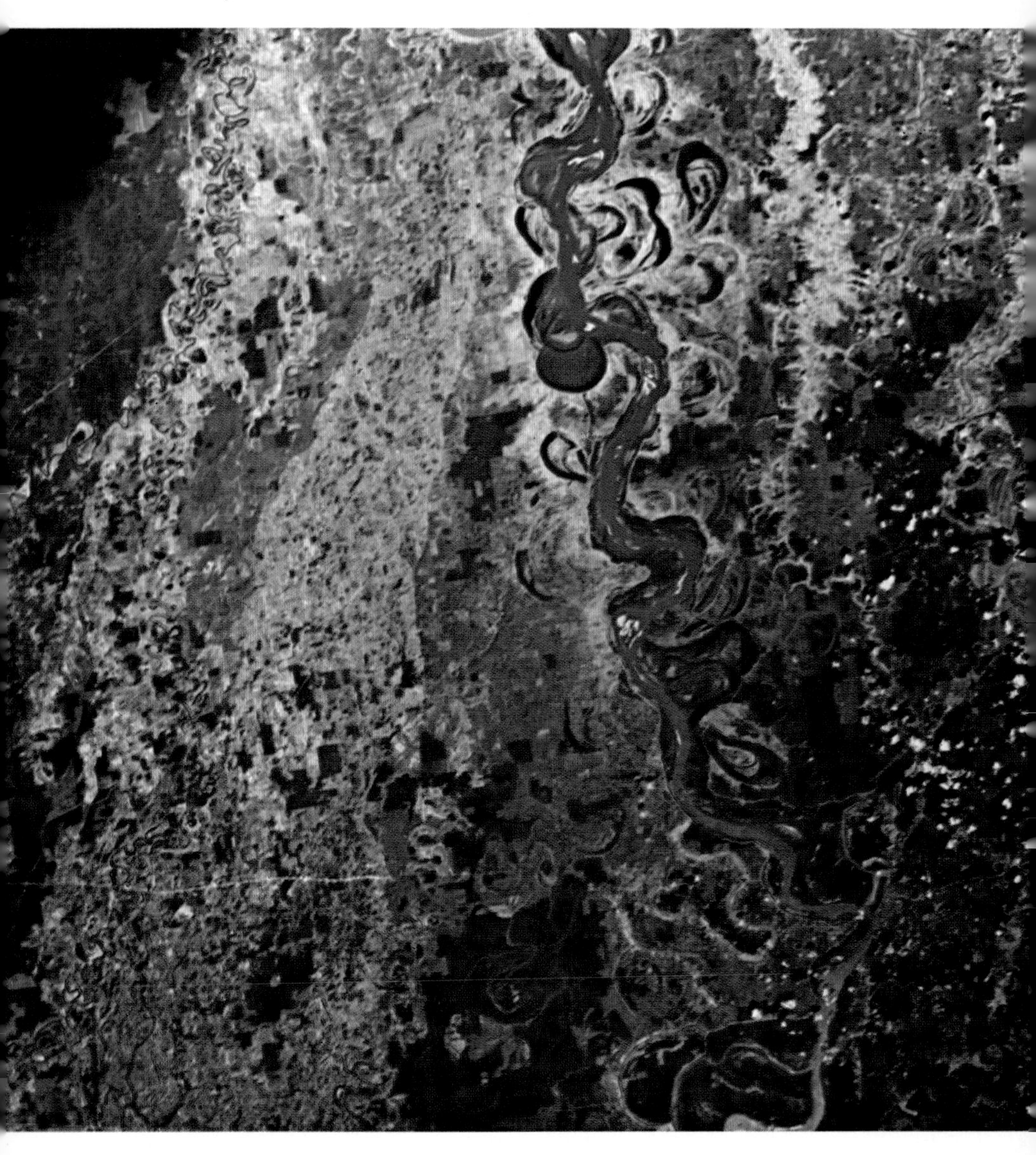

Oxbow lakes and meander scars in this Apollo 9 infrared photograph reflect recent movements of main channel of the Mississippi River near Vicksburg. Monitoring of such changes by satellite may lead to improved flood control programs. (NASA)

inputs from agricultural satellites, or else the amount and quality of information could be greatly improved for the same cost.

In many countries, currently updated statistical information on crop, livestock, and forest resources is lacking and sorely needed. Just as many nations passed from the mule or burro to the jet age in a single leap, so they may go just as quickly from hesitant, painstakingly slow ground surveys to an almost complete reliance on satellites. The major stumbling block ultimately may be not the unavailability of information but the inability, due to lack of trained personnel, to take advantage of it. User nations must develop a cadre of specialists who understand temporal, spatial, and spectral signature analysis, data retrieval and interpretation, and the means by which useful and timely information can be transferred to the farmer so that he can understand it.

One cannot consider the problems of agriculture in developing countries and ignore the role of road transport in benefiting crop production and promoting efficiency in marketing. In countries where rural roads are lacking, agriculture remains almost at a standstill; where new roads are opened, production goes up—sometimes by a factor of three. Satellites can assist in pointing out where new rural and feeder roads should be built in emerging countries, as well as aid in their maintenance. Yearly savings resulting from road construction outside the United States are estimated at $1.5 billion a year. Additional billions will come into the picture when satellite technology is called upon to help guide the construction of new roads and maintain old ones. Studies show that even minimal rural roads pay for themselves several times each year by generating greater agricultural production and reducing the costs of market distribution.

Ultimately, the agricultural planner, the commodity buyer, the land, sea, or air transporter, and dozens of other government and private interests require accurate, detailed, and current information—partly in the form of maps—on the use of land and its capability to support commercial crops, forests, livestock, natural vegetation, and wild animals, as well as the population needed to manage it. In the United States it is estimated that 60,000 separate jurisdictions determine how land is used—each with a degree of zoning authority. This has resulted in an unbelievable waste of resources, primarily because there is no overall plan of land use.

Land use patterns are caused by a multitude of complex elements, many so subtle as almost to defy identification. Even far from urban centers they are often as much influenced by economic, political, and social factors as by purely agricultural considerations, causing interactions and reverberations that can extend far beyond local or national boundaries. One nation may build up huge agricultural surpluses because it knows it has a permanent, friendly, reliable market in a habitually deficit nation. Another nation's valuable agricultural commodities may be boycotted simply because of the political philosophy of its leaders. And still another country, which theoretically should produce huge crop excesses, barely feeds its people because of archaic or unworkable agricultural management schemes. Underdeveloped nations have little knowledge of either land use or land capability, nor do they possess the surface transportation networks to exploit inaccessible, yet potentially valuable, resources. In some cases, the resources themselves are only hazily inventoried—if they are inventoried at all.

As with any great new innovation, it is difficult to foresee all or even most of the potential benefits of Earth resources satellites to the land surface of our planet. It is even more difficult to assess the full economic benefits that are likely to accrue in the years and decades to come. But it is possible to say with certainty that the application of satellite technology to geography, geology, hydrology, agriculture, forestry, and other disciplines will be revolutionary not only in developed countries like the United States but for emerging nations all over the world.

7
THE ATMOSPHERE

HUMAN LIFE AND HUMAN SOCIETY ARE inevitably influenced by the weather. Some of man's enterprises are more weather-sensitive than others: for example, summer and winter sports, agriculture, building, air and sea transportation, outdoor recreation. Office workers and train riders worry less about the weather but nevertheless can, like all people, be affected by it—violently at times.

The strongest hurricane ever to hit the United States—Camille—packed powerful winds of up to 200 mph. Before it was over, it had killed more than 300 persons and left 200,000 temporarily without homes. An estimated $1 billion in property damage was caused by the storm, which hit the Mississippi coast on 17 August 1969. What the loss of life and property might have been had not advance information on the storm been available from weather satellites is impossible to say with precision, but the Environmental Science Services Administration estimates that without such satellite information, up to 50,000 persons might have perished. Later in the year, satellite tracking of Hurricane Laurie provided the basis of forecasts that it would not strike the Gulf Coast, and hence protective and evacuation procedures were not necessary—providing savings of more than $3 mil-

Three hurricanes tracked by the Essa 5 weather satellite in mid-September 1967. Information on these hurricanes was distributed to Earth receiving stations by ATS-1. A fourth tropical storm is visible over the Pacific Ocean (left), southwest of Baja California. (ESSA)

lion on such precautions. Again, back in September 1967 when Hurricane Beulah struck, although 300,000 persons were made homeless only 41 died, thanks largely to timely warnings made possible by weather satellites. Prior to the 1970 disaster in East Pakistan, weather satellite pictures showed severe conditions brewing in the Bay of Bengal more than 10 hours before the cyclone struck. Warnings were sent to the East Pakistani center of Dacca (*moha bipod shonket*—"big danger coming") and from there to rural areas. Unfortunately the warning system was grossly inadequate, and response by the populace lethargic.

In Romania, the overflowing of the river Danube and its tributaries left some 500,000 homeless and more than 200 dead in May–June 1970, causing more property damage to that nation than did World War II. Much of the country's richest topsoil was washed away, and thousands of animals were killed.

While weather satellite data cannot stop such disasters, they can be used to prepare man for them. It is incontrovertible that enormous benefits will result from improvements in our predictive abilities. In the United States, the National Academy of Sciences has conservatively estimated a $2.5-billion annual savings from better long-range weather forecasts for the following activities: flood and storm damage prevention, new construction, fuel and electric power consumption, fruit and vegetable production, and livestock production. The academy admits that it is impossible to determine the number of lives saved around the world and the value of property safeguarded because of weather satellite information, but the number is assuredly great and the value large. It points out that until further research is undertaken, one cannot answer such questions as, "What are the quantifiable dollar benefits that would accrue to selected industrial sectors in the United States from a system that would provide a 5- to 7-day weather forecast with accuracy comparable to that of the currently available 1- to 2-day forecast over a 2-country area?" Nevertheless, the academy sees important benefits in the following areas:

1. More efficient management of the routing and scheduling of air, highway, and water traffic.
2. Decreased spoilage of perishable commodities in transit or at terminal facilities.
3. More efficient scheduling of on-site filming in the motion picture industry.
4. Improved planning of recreational and sporting activities.
5. Agriculture (for example, savings from unnecessary reseeding, fertilizing, or spraying operations; from improved timing of hay, grain, or fruit crops; or from the acceleration of harvests).
6. Construction (for example, optimum scheduling of the work force, materials, and equipment at construction sites).
7. Water management and conservation (for example, flood control —advance warning and, where possible, avoidance of damage; savings from unnecessary irrigation of vast areas).

8. Public utilities—electricity and gas (for example, more efficient methods of facility repair, maintenance, replacement, and switchover).

WEATHER SATELLITE PROGRAMS

From the point of view of man, the meteorological atmosphere extends from the surface up to 40 or 50 miles; within this envelope atmospheric mixing and local, regional, and worldwide weather patterns are developed. Most of the weather takes place in the lowest atmospheric level, the troposphere, which extends 3 to 7 miles at high and middle latitudes and nearly 10 miles at the equator. Beyond this layer is the stratosphere, a region of nearly constant temperature extending to some 30 miles that is vital to our understanding of the circulation and dynamic balance of the entire atmosphere. The mesosphere, a region of decreasing temperature, follows, ending at about 50 miles.

Meteorology is a dynamic science, concerned primarily with the study of the physics and chemistry of the lower atmosphere and with its daily and seasonal changes. Because of the strong effects meteorological processes have on the oceans and on the land, and hence on man, every possible effort is made to predict the weather and, by such predictions, to prepare for it and to modify—or even control—its phenomenal impact on society.

As with the whole atmosphere, its lowest 50 miles can be studied from the ground, in situ, and from above. Since the advent of artificial satellites in 1957, the major effort has been from above, to observe meteorological phenomena on a synoptic, repeated basis—the overview. Like the oceans, the atmosphere is ever-changing; events in one geographic area may affect those in another. Ground observations, even in populated regions, can never be pieced together with the same thoroughness and rapidity—and low cost—as from orbit. And vast expanses of ocean and land exist where no surface meteorological stations are located and from where data are obtained only occasionally, as by a passing ship or a scientific exploration party.

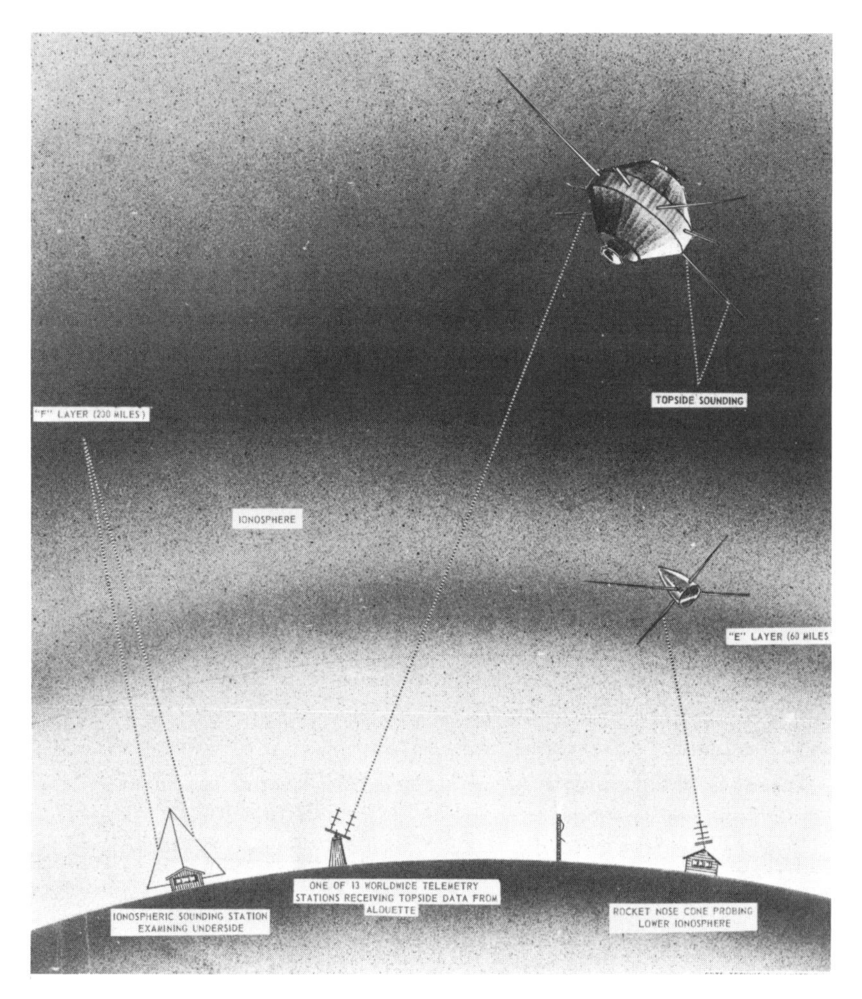

Sounding the Earth's atmosphere from "topside" with the Canadian Alouette 2, operated by the Defence Research Telecommunications Establishment. Data are telemetered to 13 receiving stations, which relay magnetic tapes to the DRTE data reduction center. Also shown are ground-based and rocket-borne sounding techniques. (NASA AND DRTE)

In order to predict the weather with any degree of certainty, large quantities of raw information must be fed into computers so that numerical analyses can be made. The atmosphere is like a huge engine, and unless data from all points of the world are available as inputs, its functioning can never really be accurately understood and significant improvement in prediction realized. Such prediction requires information on heat transfer in the atmosphere, temperatures, pressures, densities, composition, winds, rainfall, and evaporation from points scattered all over the planet Earth. Some of these data can be obtained from ground, balloon, aircraft, and rocket observations; but only artificial satellites can effectively and economically gather them on a global scale.

Weather satellites have already more than proved their worth. The United States Tiros and Essa craft alone had provided well over a million pictures by 1970. Since February 1966, the world weather has been monitored daily from space, and more than 400 major typhoons and hurricanes have been tracked. Automatic picture transmission from satellites has become so routine that commercial and military pilots regularly check such photographs before taking off on a flight. Indeed, they may soon be able to receive pictures during flight—of particular value when flying long routes where weather conditions are apt to change rapidly. While it is easy to understand the benefits of such pictures, it is difficult to place dollar values on the increased safety, reduced flying time, and added comfort they provide.

Studying the atmosphere and the weather from high altitudes began in the years following the end of World War II, when films were recovered that had been exposed in cameras carried in such rockets as the modified V-2, the Aerobee, and the Viking. Of particular interest was the discovery of a tropical storm in the Del Rio, Texas, region in October 1954—one that had not been observed by conventional methods. These flights underscored the potential of high-altitude photography, but they did not satisfy the meteorologist. To obtain regularly received synoptic inputs for use in analyses of weather, the only answer was the artificial satellite. Accordingly, with the impetus given to satellite experimentation by the International Geophysical Year (IGY), several projects involving Earth heat-budget measurements and cloud-cover mapping were carried out in 1959.

While the instruments for these early experiments were being developed during the 1957–58 IGY period, the Advanced Research Projects Agency of the U.S. Department of Defense took steps toward the creation of a satellite program directed exclusively toward general atmospheric research and the regular imaging of world weather to

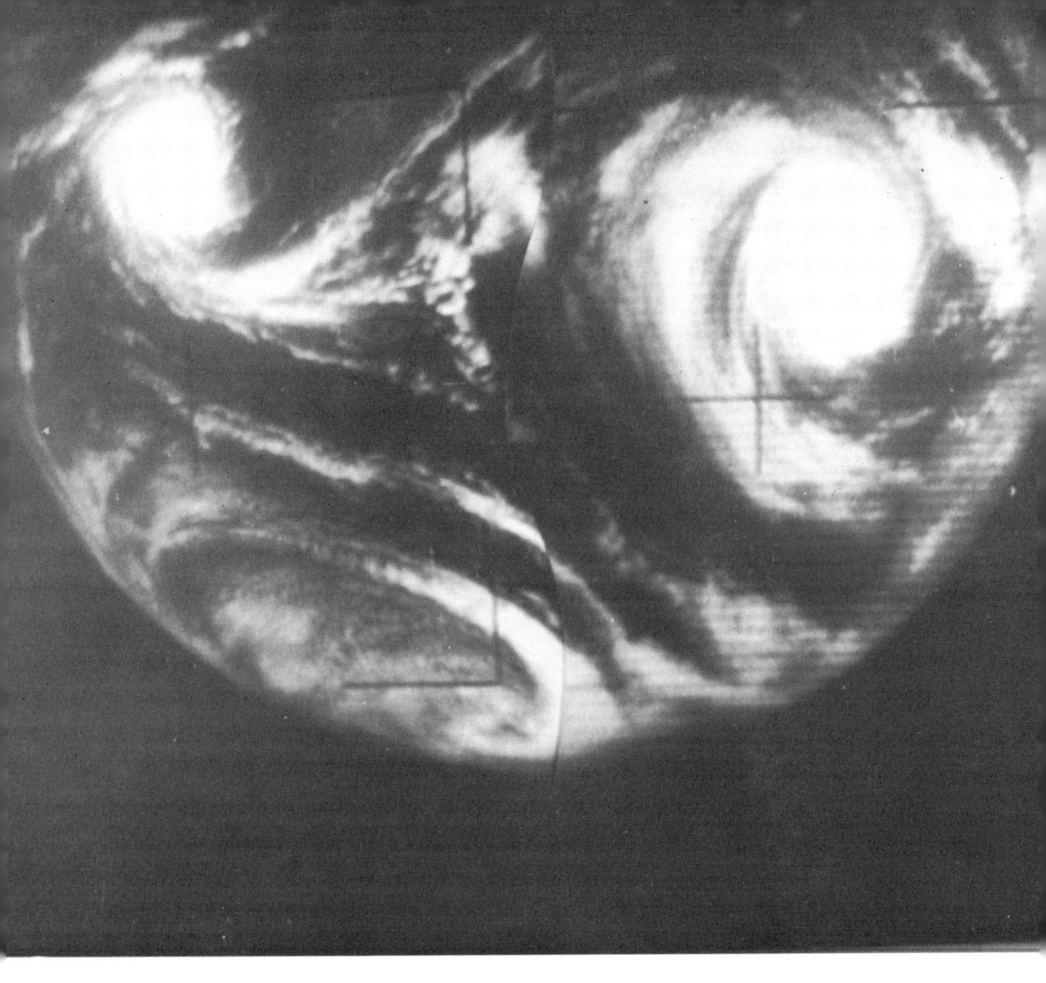

A mosaic of three atmospheric vortices in the Indian Ocean, observed by Tiros 9 in late February 1965. (NASA)

provide day-by-day forecasts. The result was Tiros (Television InfraRed Observational Satellite). In 1959 the program was transferred to the newly created National Aeronautics and Space Administration.

Between 1 April 1960 and 1 July 1965, 10 Tiros weather satellites were launched. They had useful lifetimes ranging from 89 to 1,809 days; each took between 22,952 and 125,331 pictures. Tiros 1 through 8 were virtually identical. They weighed from 270 to 300 pounds, were built in the form of an 18-sided polygon of aluminum alloy and stainless steel, and measured 42 inches in diameter and 22 inches in height. Power was supplied by 9,200 solar cells. Within nine days of its launch, Tiros 1 had located a typhoon 800 miles east of Brisbane, Australia.

TABLE I
TIROS SATELLITE PROGRAM

FLIGHT NO.	DATE OF LAUNCH	USEFUL LIFE (days)	PICTURES TRANSMITTED
1	1 April 1960	89	22,952
2	23 November 1960	376	36,156
3	12 July 1961	230	35,033
4	8 February 1962	161	32,593
5	19 June 1962	321	58,226
6	18 September 1962	389	68,557
7	19 June 1963	1,809	125,331
8	21 December 1963	1,287	102,463
9	22 January 1965	1,238	88,892
10	2 July 1965	730	79,874

Tiros 8 was equipped with not only a regular wide-angle camera but a new type of camera integrated into the Automatic Picture Transmission (APT) system. With this system, pictures could be automatically transmitted to the Earth as the satellite orbited above; received by many small, inexpensive ground stations, instead of a few major command and data acquisition centers; and read out through a facsimile machine. The maximum slant range from satellite to APT station is 2,000 miles. When received by local forecasters, the pictures represent conditions having occurred a mere 3½ minutes earlier. The APT system permits weather data to be received at low cost by some 500 receivers in more than 50 nations all over the world (in some areas, APT represents the sole weather-forecasting facility).

Whereas the sensors of the first eight Tiros satellites were located at the base of the spacecraft, in Tiros 9 and 10 they were placed on the rim, 180 degrees apart. This change was made because of serious limitations in the earlier models. These, because of their orientation, could monitor only 10 to 25 per cent of the Earth's surface in a single day. Moreover, since their sensors did not always point straight down, interpreters were forced to spend much time in orienting, identifying, and rectifying the photos received. In the new system, the satellite was turned on its side in orbit so that it rolled, pointing sequentially

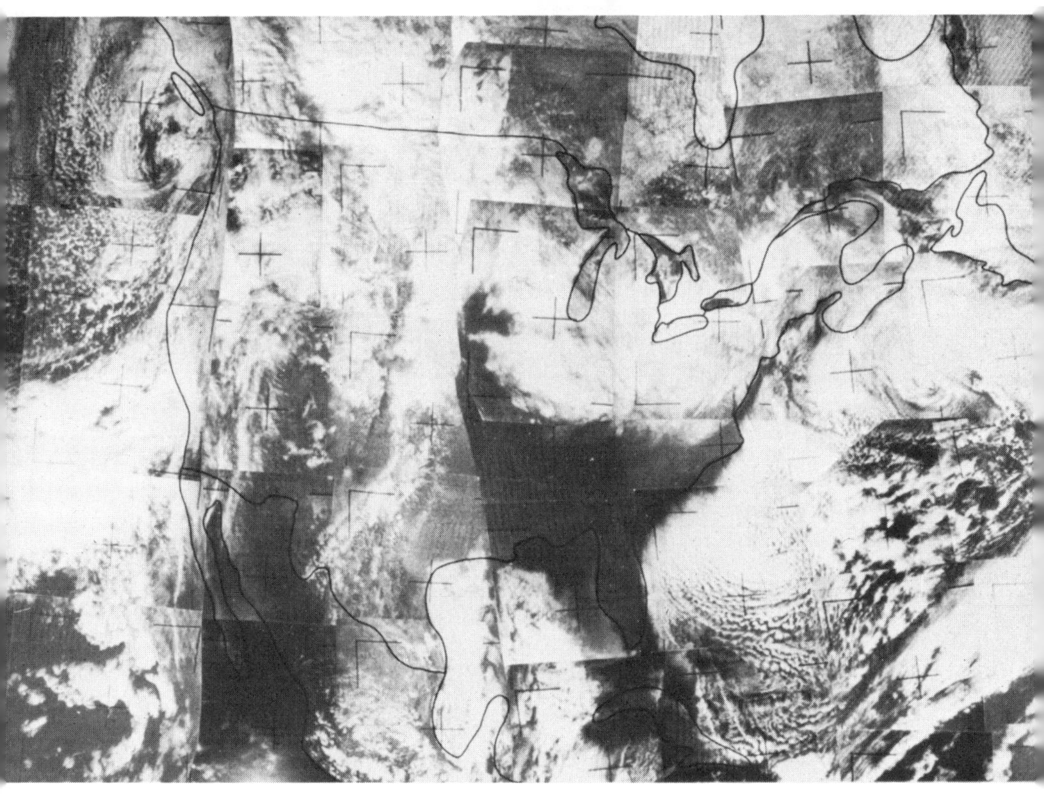

First complete coverage of North America by Essa 1, showing heavy-rain-producing storm along the West Coast, ice in Lake Erie, snow cover over the northeastern United States, and clouds in a cold airflow off the East Coast. (ESSA)

first one camera and then the other directly toward the surface below. Moreover, these last Tiros satellites were inserted into polar orbits rather than given the 48- and 58-degree inclinations of earlier members of the series. Since the Sun was therefore high in the sky behind the spacecraft, all the sunlit surface below could be photographed each day.

The Environmental Science Services Administration used design and polar-orbit features of Tiros 9 in its series of nine Essa (Environmental Survey SAtellites) spacecraft, all launched within the Tiros Operational Satellite (TOS) program. The first in the series, Essa 1, was orbited on 3 February 1966; the last, Essa 9, on 26 February 1969, a little over three years later. These craft helped to supply the National Operational Meteorological Satellite System with daily cloud-cover and other data.

Two types of Essa satellites are used in the TOS system: those with even numbers, carrying APT equipment; and those with odd numbers, carrying two advanced, 1-inch vidicon cameras, or AVCS (Advanced Vidicon Camera System), plus two cross-connected tape recorders on each of which 48 frames can be stored. The pictures

TABLE II

ESSA/NOAA SATELLITE PROGRAM

FLIGHT NO.	SENSOR SYSTEM	DATE OF LAUNCH	USEFUL LIFE (days)	PICTURES TRANSMITTED
Essa 1	AVCS	3 February 1966	861	111,144
Essa 2	APT	28 February 1966	1,461 [a]	125,288 [a]
Essa 3	AVCS	2 October 1966	241	92,076
Essa 4	APT	26 January 1967	110	27,129
Essa 5	AVCS	20 April 1967	1,037 [a]	86,715 [a]
Essa 6	APT	10 November 1967	725 [a]	64,154 [a]
Essa 7	AVCS	16 August 1968	338	39,953
Essa 8	APT	15 December 1968	974 [a]	95,000 + [a]
Essa 9	AVCS	26 February 1969	900 [a]	81,000 + [a]
Itos 1 [b]	AVCS APT SR	23 January 1970	509	107,140 (+ 11,820 hrs of SR data)
Noaa 1 [b]	AVCS APT SR	11 December 1970	210	25,222 (+ 6,000 hrs of SR data)

[a] Through 15 August 1971, at which time the satellite was still operating.
[b] Following Essa 9, the APT camera system and AVCS were combined aboard a single spacecraft—Itos (Improved Tiros Operational Satellite). Later, with the renaming of the Environmental Science Services Administration as the National Oceanic and Atmospheric Administration, the designation was changed to Noaa. The next polar orbiter, expected to be launched in late 1971, will be Noaa 2. The Itos/Noaa design carries a Scanning Radiometer (SR) as well as a solar proton monitor.

from the APT-fitted satellites are scanned on board and transmitted to APT receiving stations on the ground. As soon as a given picture has been scanned and transmitted, the vidicon is erased and made ready for the next picture. As for the AVCS stored-data satellites, pictures are taken every 260 seconds when the craft is in daylight.

Digital mosaic of cloud pictures taken by Essa satellites over the South Pole. Image signals from the satellites are sampled at short intervals and converted to numbers representing gradations of brightness. The computer locates each digital value at precisely the right geographic location; then a computer tape containing this information is displayed on a kinescope and photographed, with the above result. (ESSA)

Each image is retained by the photosensitive layer on the face of the vidicon tube for 7 seconds, during which time the picture is converted to signals that are stored on the magnetic tape. These signals are later transmitted to central command and data acquisition stations at Fairbanks, Alaska, and Wallops Island, Virginia. From there they are sent on to the National Environmental Satellite Center at Suitland, Maryland, where they are converted to numbers representing gradations of brightness in the picture images. Computers transfer the brightness values onto map projections, resulting in digital mosaics. These are sent to weather stations all over the United States and, from the World Meteorological Center in Washington, to recipients around the world by radio-facsimile broadcast—except in the case of Russia, which receives its data over a direct circuit into another World Meteorological Center in Moscow. Stored-data Essa satellites also carry arrays of radiation sensors to gather information on the Earth's heat balance; these measure solar energy reflected by the surface and by the atmosphere as well as long-wave radiant energy emitted by the Earth.

The third phase of the original TOS program is the Itos (Improved Tiros Operational Satellite) series. As compared with the Essa series, Itos satellites are designed to provide double the daily weather coverage, at lower cost and more efficiently. Also their operational lifetimes are longer, and attitude stabilization is much improved. And the newer satellites are able to take infrared photos of cloud cover at night and to measure cloud-top and Earth surface temperatures, unlike the earlier series. Itos, in effect, took some of the sensors then being developed by Nimbus and put them into operational service.

Itos I was orbited on 17 January 1970; five months later it was declared operational by NASA and turned over to the Environmental Science Services Administration. Weighing 682 pounds—more than twice the weight of the Essa series of satellites—and of rectangular instead of hatbox shape, it carried *both* APT and AVCS, plus scanning infrared radiometers, into its near-polar, 890- to 918-mile high orbit around the Earth. A complete APT picture sequence contains 11 pictures taken at 260-second intervals. When the programmed sequence is over, the pictures are transmitted to the APT field stations via real-time data link. The wide-angle, high-resolution AVCS picture sequence lasts about 48 minutes, during which time 11 frames are taken at 260-second intervals and immediately stored. At the time the last frame is taken and recorded, the AVCS is temporarily deactivated pending the start of a new sequence. The radiometer data from the 0.52- to 0.73-micron visible region (day) and the 10.5- to 12.5-micron infrared region (day and night) are transmitted in real time (as events occur) to local stations and in delayed time (magnetically stored) to command and data acquisition stations.

As Tiros was progressing toward operational status, experimental equipment was being tested in another satellite series, Nimbus. Unlike Tiros, it is Earth (rather than space) stabilized and oriented, permitting greater geographical coverage. Placed in near-polar, Sun-synchronous orbits, Nimbus employs its orbital movement to secure wide-latitude coverage, and the Earth's rotation to assure wide-longitude coverage.

Carrying APT and AVCS features plus a radiometer, Nimbus 1 was launched on 28 August 1964 and commenced providing both day and night cloud pictures. After about a month in orbit the solar paddle drive system failed; since the paddles could not rotate, a power drain resulted and Nimbus 1 went out of operation. During the course of its 380-orbit lifetime it located and tracked many hurricanes and typhoons and provided the first nighttime pictures of cloud cover. In all, it returned more than 27,000 day and night pictures of the Earth and its cloud cover.

The 912-pound Nimbus 2, orbited on 15 May 1966, contained the same sensors as its predecessor plus equipment to measure the percentage of solar radiation reflected by the Earth back out through the atmosphere. It operated far beyond its 6-month design lifetime, being shut off in mid-January 1969—after 32 months—because of tumbling problems. Some 112,500 photos were transmitted by the advanced vidicon cameras, while the APT provided more than 300 ground stations with pictures over an 8,000-hour period.

The first attempt to put up the third Nimbus, on 18 May 1968, was a failure. Accordingly a repeat satellite was built and subsequently orbited on 14 April 1969. Designated Nimbus 3, this 1,269-pound spacecraft carried seven meteorological experimental packages, including an infrared interferometer spectrometer, capable of measuring infrared radiation at various layers in the atmosphere and also cloud-top or Earth surface temperatures (measurements are later processed mathematically to yield vertical temperature profiles). It was also possible to ascertain the presence of such constituents as carbon dioxide, nitrous oxide, and methane. Aboard was a new interrogation recording and location system, designed to show that a satellite could pinpoint the positions of meteorological recording platforms placed in balloons, buoys at sea, and ice islands. By 1 March 1970, Nimbus 3 had transmitted more than 150,000 pictures and had measured the atmospheric temperature and water vapor profiles down to the surface for the first time—a major breakthrough. It was still operating in a highly satisfactory manner in 1971.

TABLE III
NIMBUS 4 METEOROLOGICAL EXPERIMENTS

DEVICE	SPECTRAL BANDS (microns)	PRIMARY PURPOSE OF EXPERIMENT
Temperature-humidity infrared radiometer	10.5–12.5	Daytime and nighttime surface and cloud-top temperature measurements; cloud mapping
	6.5–7.0	Atmospheric water vapor mapping
Infrared interferometer spectrometer	8–20	Measurement of atmospheric temperature profile, ozone, water vapor, surface temperatures, and minor atmospheric gases
Satellite infrared spectrometer	11	Surface and cloud-top temperature measurements
	13–15	Atmospheric temperature profile measurements
	19–36	Atmospheric humidity profile measurements
Ultraviolet solar energy monitor	0.12 0.16 0.18 0.21 0.26	Monitoring of changes in solar radiation

TABLE III (continued)
NIMBUS 4 METEOROLOGICAL EXPERIMENTS

DEVICE	SPECTRAL BANDS (microns)	PRIMARY PURPOSE OF EXPERIMENT
Selective chopper radiometer	13–15	Atmospheric temperature profile measurements
Filter wedge spectrometer	1.2–2.4 3.2–6.4	Atmospheric water vapor measurements
Backscatter ultraviolet spectrometer	0.25–0.34	Determination of atmospheric ozone distribution
Image dissector camera system	0.45–0.65	Daytime cloud mapping
Interrogation, recording, and location system	—	Data collection from platforms

Adapted from *The Nimbus IV User's Guide*. Greenbelt, Md., 1970: Goddard Space Flight Center, National Aeronautics and Space Administration.

Nimbus 4 entered a near-polar, 107.36-minute orbit, some 690 miles above the Earth, after a successful launch from the Western Test Range in California on 8 April 1970. The objectives of this 1,360-pound satellite were to study the spatial and temporal distribution of atmospheric structure (especially heat, ozone, and water vapor) and to determine variations in solar ultraviolet radiation. Of the nine experiments, five were improved versions of earlier procedures and four were completely new.

The Application Technology Satellite (ATS) program is not primarily meteorological, but from positions in geosynchronous orbits these craft can conduct the observations needed to track short-duration storms and to follow the motions of the atmosphere as revealed by clouds. Depending on the payload incorporated in the 56-inch-diameter satellite, the weight varies between 650 and 790 pounds. Both ATS-1 and ATS-3 carried spin-scan cloud cameras into orbit, in December 1966 and November 1967 respectively, the first over the Pacific Ocean and the second over the Atlantic. Meteorologists,

Considered the highest-quality photograph received from a United States weather satellite up to May 1970, this Nimbus 4 view shows snow-covered Scandinavia, the partially ice-covered Gulf of Bothnia, and northern cloud patterns associated with a storm in the Barents Sea. (NASA)

receiving photos of the entire disk of the Earth every 20 minutes, are able to sequence them and prepare time-lapse motion pictures of changing cloud patterns. Such films are employed as an aid to determining the development and progress of hurricanes and other major storms. Moreover, daily film loops supply wind information which is fed into the numerical analysis programs of the National Meteorological Center. Data processed from an ATS are transmitted through the WEFAX (WEather FAcsimile eXperiment) system for use in the United States and more than 50 other countries.

Excellent color photographs of meteorological phenomena have been taken by the Mercury, Gemini, and Apollo manned spacecraft; by a number of nonmeteorological satellites, such as Dodge—which on 25 July 1967 took the first color picture of the full Earth; by lunar probes—Lunar Orbiter 1 took the first photograph of the Earth from the vicinity of the Moon on 23 August 1966 (even from there, cloud cover patterns were clearly evident); and by unmanned rockets and carrier vehicles—Saturn 5 flight AS-501 took excellent color photos. The Russians have secured similar photos from manned satellites in the Vostok, Voshkod, and Soyuz series; from unmanned meteorological satellites in the Kosmos (including Nos. 122, 144, 184, and 243) and Meteor series, and occasionally from other craft such as the Molniya I series, which have taken continentwide pictures at a variety of magnifications.

SENSING FROM REMOTE SURFACE AND AERIAL PLATFORMS

Since so much of the world's weather occurs over the oceans and remote land areas, the employment of remote-sensing platforms that can be monitored and interrogated by synchronous-orbit meteorological data relay satellites is particularly attractive. Superpressurized "horizontal-sounding" balloons, designed to float with (and measure) wind currents at constant density levels, are fitted with sensors to monitor temperature, pressure, and water vapor. Normally they transmit their accumulated data upon

TABLE IV
CATEGORIES OF METEOROLOGICAL OBSERVATION

MAJOR ASPECT	MEASURABLE PHENOMENA [a]
Vertical structure of atmosphere	Density, temperature, pressure, winds (tropospheric and stratospheric), clouds, water vapor, ozone
Surface conditions	Temperature, winds (over oceans, determined from observation of sea state), ice and snow cover, soil moisture, frozen ground, roughness of ground
Special phenomena	Jet stream and clear-air turbulence, squall lines and tornadoes, hurricanes (cloud structure, temperature, pressure, radiation)
Clouds	Identification of types, global census and degree of cover, climatic effects, organization and development, ice or water formation, altitudes of tops
Atmospheric electricity	Lightning (horizontal length, wide-band sferics, upward strokes), electric fields at flight level
Radiation	Solar constant, albedo (angular variations)
Precipitation	Present and past fall, intensity, global geographical distribution
Atmospheric constituents and aerosols	Type (e.g., carbon dioxide, sulfur dioxide, carbon monoxide), turbidity, pluming

[a] Partially through the use of interrogation and relay of data from surface and airborne observation platforms.

command, but they could dump them at prearranged intervals. A program known as GHOST (Global HOrizontal Sounding Technique) also has been created. GHOST balloons have circled the world many times, providing invaluable data on winds over inaccessible stretches of ocean and land. At sea, ocean buoys—typified by the Navy Nomad program—measure surface weather data as well as oceanographic parameters. Weather equipment is also placed on naval, coast guard, and commercial ships that can be tied into relay satellites.

Two systems for locating remote free-floating balloons and fixed and free-floating surface platforms have been developed: the IRLS (Interrogation, Recording, and Location System), employed with

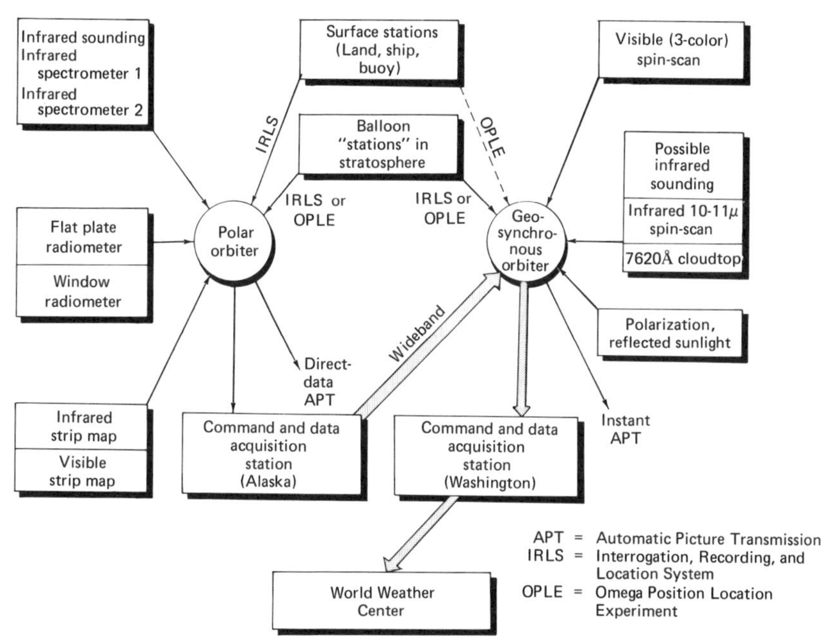

Interlocking polar- and geosynchronous-orbit meteorological satellite systems mesh their observations with those obtained from balloons and surface stations to produce data for the World Weather Center in Washington, D.C. (RICHARD P. MCKENNA)

Nimbus satellites; and the OPLE (Omega Position Location System), tied to the ATS program. The major constraint in any mixed space-Earth observation system is the development and packaging of the electronics, particularly for superpressurized balloons. To avoid posing a hazard to aviation, the electronics and power supply together have to be very light and small, especially in consideration of the fact that up to 2,000 balloons may be involved.

Among the concepts for future atmospheric research is a radio occultation experiment involving a master satellite and a number of coorbital slave satellites, designed to shed further light on the density of air at many altitudes. Atmospheric density is basic to the understanding of local and regional conditions and to predictions of future meteorological activity. The idea behind the experiment is to pass a spinning microwave beam across different paths through the atmosphere, using the refraction of the beam to yield atmospheric density values. A system involving one master Radiating Source (RS) satellite, one Receiver-Recorder (RR) satellite, and eight sequentially spaced Receiver-Transmitter (RT) satellites has been recommended by the Environmental Science Services Administration. As the detected signal is sensed by an RT satellite, the event is transmitted to the RR satellite, which collects all data for relay to ground stations. The time required for the signal to reach the RT satellite is determined by the density of the atmosphere through which the signal passes—being a function of the refractive index change along the RS–RT path. By using eight RT satellites, vertical density sounding is possible. Density could also be determined by measuring phase variations of radio waves traversing tangentially the atmosphere.

ORBITAL DETECTION AND MONITORING OF ATMOSPHERIC POLLUTION

The detection, monitoring, policing, and control of atmospheric pollutants comprise one of the greatest challenges to mankind. Since the latter part of the nineteenth century, man has carelessly, blindly, and on a terrifying scale discarded contaminants into what he thought was a boundless dump. Today, he realizes that he may be doing irreparable damage to the ocean of air that he and other products of billions of years of evolution share. Man's prowess in space may prove to be a decisive weapon in the war against atmospheric pollution, a war that may save his species from extinction.

In the broadest terms, one is interested in knowing the sources of pollution, its local concentration and regional diffusion, the nature of individual contaminants, and the effects they may have on climate in general and weather in particular. Global reconnaissance of atmospheric contamination will ultimately be possible, using a system of instrumented artificial satellites. Two principal areas of investigation exist: (1) at and just above the surface-air interface, and (2) the upper troposphere and stratosphere. Moreover, two time scales are of interest: (1) the present and immediate future (up to a month), and (2) the long-term future (months, years, even centuries).

Lower-atmosphere detection is complicated by the fact that pollution generally concentrates in thin layers close to the ground rather than rising and mixing at higher levels. This condition means that sensors will have to be sensitive to low-layer contaminants as distinct from other substances found along the satellite-to-ground observation path. In order to monitor effectively the sources of air pollution, the intensity and movement of the effluents, and the nature of the individual constituents, a combination of satellite and ground detection systems will be necessary. On the information they provide, maps can be constructed showing pollution flow from heavily populated and industrial areas on a regular (once or twice a day) basis. Pollution-monitoring satellites will not only employ their own sensors but also act as relays for data transmitted up from the ground stations. Such satellites will eventually serve policing and predictive functions as well as providing detection and reconnaissance. Based on satellite-observed weather patterns, predictions of potential pollution accumulations and their density will be invaluable in initiating preventive measures, including weather modification, shutdown of industrial plants, and control of motorcar and truck traffic.

On a broader and longer-term scale, satellites will be of vital importance in providing surveillance of the global buildup of atmospheric pollution, not only in the lower atmosphere but in the upper troposphere and the stratosphere. Realizing, for example, that the increase of carbon dioxide would tend to cause rises in worldwide temperatures, while the increase of particulate matter suspended in the air would tend to lower temperatures, meteorologists will want to determine what long-range effects man is having on his climate—and what permanent, perhaps irreversible, damage he is doing.

Before air pollution satellites can become a reality, much sensor research must be undertaken. Individual sensors must be able to identify major contaminants in the atmosphere. High-resolution absorption spectroscopy and ultraviolet laser techniques appear particularly promising. Studies of direct emissions of atmospheric pollutants also

Pollution can be measured easily from the ground, but only local coverage is possible. For a worldwide detection, monitoring, and policing program against air pollution, the satellite offers the only economically attractive solution. Here, pollution is observed causing chlorotic dwarf symptoms among white pines. (U.S. DEPARTMENT OF AGRICULTURE)

bear close watching. An important step toward the development of atmospheric pollution sensors was taken by NASA's Langley Research Center when it initiated a development program for a spaceborne carbon monoxide sensor. It is estimated that some 500 million tons of carbon monoxide is suspended in the atmosphere at any one time, with motorcar exhaust, industrial plant emissions, and other human activities generating 200 million tons each year. One of the purposes of the NASA experiment is to determine how carbon monoxide is removed from the air. Another is to learn where the poisonous gas is concentrated.

Many profound problems remain to be solved before an effective attack on global pollution can be undertaken. Some of these involve basic atmospheric processes whose nature is becoming understood on the basis of meteorological-satellite information. Others are not so clear. What, for example, are the interactions between pollutant emissions and meteorological processes, including air flow and stratification, turbulence, humidity, rainfall, and sunlight? What are the primary mechanisms of pollutant transport on local, regional, and global scales? How are dispersion rates influenced? What photochemical, chemical, and physical changes do pollutants undergo once they have been injected into the air? How does mixing of pollutants occur, particularly between the northern and southern hemispheres and between the troposphere and the lower stratosphere? What is the role of the stratosphere in storing pollutants, and when, how, and to what extent are they returned to the troposphere? Are pollutants causing irreversible changes in the Earth's heat balance and overall climate? With the knowledge that suspended particles diffuse and absorb incoming radiations from the Sun and outgoing infrared radiations from the Earth, what role have these aerosols played in the heating and cooling of the atmosphere? Climatologists suspect that major climatic changes reflected in geologic history were triggered by small (a few degrees) worldwide temperature fluctuations. If pollution can cause changes of this order, man would have to react rapidly and decisively before it was too late.

Many studies are urgently needed to shed light on such diverse questions as how much working time is lost to a nation each year because of workers' inability to produce under the influence of air pollution, and what is the meaning to a society of shortened life spans and increased illness due to air pollution in urban and industrial centers. Other concerns include the esthetic and psychological implications, the extent of damage to plants and animals, and the insidious destruction of the artistic works of man—exemplified by Venice, Italy, where some of the greatest creations in history are disintegrating at an alarming rate because of effluents from nearby industries. The only practical method of detecting and monitoring pollution on a regular and worldwide basis, and of observing its deleterious effect on the environment and the works of man, is by satellite. There is no economical substitute for the continuous overview.

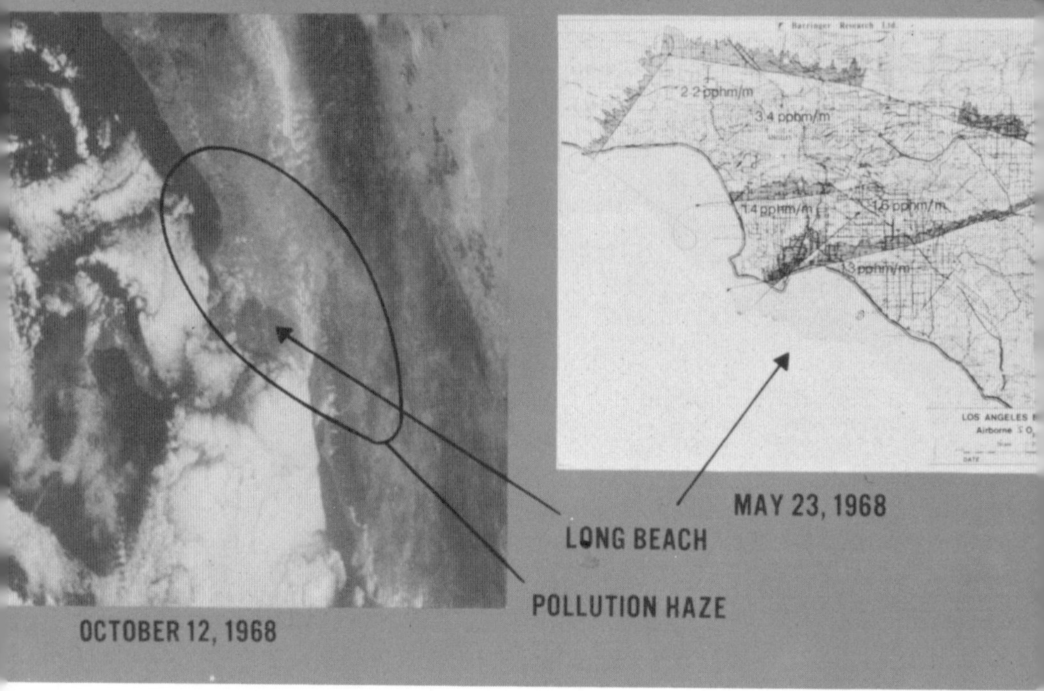

Pollution haze observed over the Los Angeles area by the Apollo 7 satellite (left) and by an aircraft survey of sulfur dioxide. (NASA)

WEATHER MODIFICATION

The World Meteorological Organization, individual national weather research agencies, and private interests continue to investigate the feasibility of producing the best possible weather for the greatest number of people by modifying existing conditions. Researchers recognize that, because of the enormous energies involved in atmospheric processes, no immediate thought can be given to the direct application of counterenergies. Rather, they direct their attention to triggering mechanisms that take advantage of instabilities inherent in the atmosphere. These include the phase instability of water, the colloidal instability of cloud particles, the convective instability of the overall atmosphere, and the baroclinic instability of large-scale circulation.

Triggering actions could cause hurricanes to release their energy before reaching inhabited areas or prevent clouds from releasing it at the wrong time. Such actions could dissipate supercooled fog and eliminate warm fog from airport runways. Modification of convective cloud patterns could increase snowfall in mountain areas where winter sports are under way. Artificial creation of cirrus clouds could change the temperature by modifying the radiation balance at particular locations. Other goals of interest to meteorologists are hail suppression, lightning suppression, the modification of large-scale circulation patterns, and the changing of the microclimates of certain plants.

A typical modification now made with some success is the seeding of supercooled liquid clouds with dry ice and silver iodide to produce rain. This technique could be materially aided by cloud surveillance, using satellites to pick out "targets" over desert lands. However, scientists must make thorough studies of the possible consequences of weather modification before man begins to tamper with the weather beyond such things as local cloud-seeding and fog dissipation.

The principal contributions of the space effort to weather modification are (1) to assist in predicting and locating such weather situations as appear to be potentially dangerous—and susceptible to modification; (2) to provide inputs upon which assessments can be made of the degree of success of the modification attempts; (3) to help monitor artificially induced changes in the weather that were not intended; (4) to provide long-term surveillance of atmospheric changes due to concentrations of pollutants; (5) possibly to furnish launch platforms from which rockets loaded with seeding agents could be fired down into the atmosphere; and (6) possibly either to concentrate or to interfere with incoming solar radiation over a local area.

In 1969, Hurricane Debbie—spotted early by satellites—was noticeably weakened by silver iodide seeding. Several hours after initial seeding on 18 August, winds dropped 30 per cent, from 98 to 68 knots. To provide a control, no seeding was attempted on the 19th, with the result that the storm reintensified. Further seeding took place on the 20th, when 99-knot winds were brought down to 84 knots—a 15 per cent improvement.

MANNED SPACECRAFT METEOROLOGICAL OBSERVATIONS

The role of man in orbit can provide invaluable support to the continued development of satellite meteorology. The most obvious advantage is the ability of the astronaut to exploit transitory events or observational phenomena of immediate interest that were not predicted, for which no specific procedure or sequence of investigation had been planned, and which might be missed by unmanned satellites. This is exemplified by the Apollo 9 Earth-orbital flight of March 1969. As their spacecraft passed over an area where pictures received from Essa 8 and ATS-3 had suggested to meteorologists that tornado conditions might be developing, the astronauts conducted careful surveys and reported to NASA ground personnel what they had seen. Coupled with information from other sources, their reports became part of a Weather Bureau tornado alert for the inhabitants of the potentially affected area.

Man in space can also provide important functions in monitoring diverse experiments; testing early-model, perhaps bulky, instrumentation not yet ready—and perhaps unsuitable—for unmanned satellites; repairing faulty instrumentation; and modifying the observational program as a result either of his own analysis or of requests from Earth-bound scientists with whom he is in regular communication.

THE OUTER ATMOSPHERE AND SPACE DISTURBANCES

Beyond the troposphere—the lowest atmospheric level, where man lives and where most of the weather takes place—and the stratosphere are the mesosphere, the thermosphere, and finally the exosphere. The mesosphere and thermosphere are characterized, respectively, by a lowering of temperature to about −140°F, then a rise in temperature out to a distance of a little more than 80 miles, where the thermosphere ends. Finally comes the exosphere, where the few air molecules that are left occasionally escape into space. Although there is no true end of the atmosphere, molecules are very hard to detect farther than the 400- to 600-mile region. Beyond the exosphere is what is known as the magnetosphere, a region of intense radiation of solar origin trapped in the Earth's magnetic field. Within this region are two zones of concentration termed the Van Allen Radiation Belts, after their discoverer. The inner belt consists principally of protons and extends from 825 to nearly 2,500 miles beyond the Earth, while the outer belt is made up largely of electrons and lies from 8,000 to 12,000 miles out, on the average—although at the equator it may extend to nearly 40,000 miles.

Within the atmosphere itself are other regions that coincide to a greater or lesser extent with the zones already described. One is the chemosphere, or ozone layer, which has a peak ozone concentration at 19 miles altitude. Continuously created by chemical processes, it absorbs most of the lethal solar ultraviolet radiation, thereby protecting life at the surface. A second zone is the ionosphere, or region of ionization. It results from the action of impinging solar radiation on the nitrogen and oxygen in the atmosphere, forming positively charged ions of nitric oxide and oxygen plus free electrons. Depending on concentration and altitude, a number of layers are identified: the D layer (40 to 55 miles), the E layer (55 to 90 miles), and the F layer (90 to about 250 miles). These ionized layers are able to reflect radio waves, and hence make possible radio communications around the Earth. Occasionally the ionosphere is disturbed by solar radiation pulses, creating ionospheric storms which result in the ejection of excess electrons. The electrons in turn absorb some of the energy of radio waves, causing radio fadeouts. Another effect of increased solar activity is the spectacular light displays, or auroras, in the polar regions, which are caused by highly charged particles accumulating at the Earth's north and south magnetic poles.

The outer atmosphere is important to man partly because of its role in communications, partly because it is a major arena for solar-terrestrial interactions, and partly because it is an environmental region through which unmanned and manned satellites orbit and extremely high altitude aircraft fly. This part of the atmosphere is influenced by the continuous bombardment not only of energetic solar particles

—the electrons and protons already mentioned—but also of X rays and ultraviolet light.

Streams of charged particles from the Sun affect not only radio telecommunications on Earth but also the safety of man in the outer atmosphere and space and the reliability of instrumentation in unmanned satellites. In effect, these phenomena of the outer atmosphere

Wire mesh satellite (OV1-8) inflates to 30 feet in diameter after being launched into space. The satellite is designed to study pressure of solar rays, particle drag, rigidity of the wire structure, and its potential in a passive communications relay system. (U.S. AIR FORCE)

and nearer reaches of space are analogous to lower atmospheric weather, and can thus be thought of as "space weather." Any disturbance of this weather is a "space disturbance." And knowledge about, and warnings concerning, space weather constitutes a definite and valuable space dividend.

In order to conduct the desired research, an integrated program of atmospheric and solar studies is required. This includes the interaction of solar protons with the upper atmosphere, the interaction of the solar wind with the magnetosphere, and the energy of particles trapped in the Earth's magnetic field. Other factors are auroral disturbances and ionospheric storms, general solar activity and solar-terrestrial relationships, and the theory of magnetic storms. Many of the data needed for making space disturbance forecasts are gathered from instrumented satellites. Such forecasts are needed not only for vehicles, like Apollo and Skylab, traveling beyond the sensible atmosphere, but also for supersonic aircraft flying high enough so that interactions of solar flare energy and the atmosphere could pose possible hazards to passengers and crews.

Various spacecraft of NASA (Pioneers, Explorers, ATS vehicles, and orbiting solar observatories) and of the Department of Defense (particularly Velas and Solrads) provide regular monitoring of solar flare activity. In the future, specially designed satellites for monitoring space disturbances may be placed in polar orbits that are Sun-synchronous along the dawn-dusk line, and not at 45 degrees to this line as is the presently aloft Itos 1, with its solar proton monitor aboard. Such orbits would provide a continuous opportunity for monitoring solar proton arrivals in the upper atmosphere. Other monitors could be lofted into Earth-synchronous orbits, from which continuous, real-time communications would be possible with ground stations. Since an Earth-synchronous satellite is occulted by the Earth for approximately 110 minutes each day, two satellites could be orbited to provide continuous monitoring of the protons. Such a system would be particularly valuable for supersonic transports, which could not tolerate the delays inherent in the mode of storing and relaying data by low-altitude polar orbiters.

The study of the effects of ionization on physical and chemical processes in the upper atmosphere is called aeronomy. This science also encompasses studies of auroras, the temperatures of electrons and ions, ion densities, and natural radio signals occurring in ionized media. Ionospheric phenomena are determined by interactions with the radiation environment of the magnetosphere on the one hand and of the lower atmosphere on the other.

Aeronomic studies can yield dividends in the improvement of communications techniques. Experiments in the lower ionospheric region can be made with ground-based radio ionosondes, with atmospheric sounding rockets, and with small rockets retrofired from satellites. Studies that can be made in the lower ionosphere include those of ionospheric currents and of the molecular constituents. Both affect our understanding of solar-terrestrial relationships, communications, and magnetic disturbances.

So-called "topside sounders," or satellites traveling at altitudes of about 600 miles, are employed to probe the electron density of the ionosphere from above. Other satellites map the ionosphere from within, orbiting at altitudes between 120 to 200 miles. Among the ionospheric sounders are the Alouette series of satellites, Explorer 20, and Isis 1. Still additional vehicles conduct natural and man-made radio noise and interference research, so that Earthwide maps can be constructed showing radio-noise contours at satellite altitudes. Included aboard these craft are devices to detect lightning discharges all over the world. The economic benefits that derive from research in satellite aeronomy are obvious when we consider that more than $20 billion a year is spent by the telecommunications industry in the United States.

After more than a decade of continuous meteorological satellite observation, today United States craft alone photograph and take daily soundings of the atmosphere over about half the surface of the Earth. The resulting data are put to use in almost every area of activity that is affected by or is dependent on the weather. Using satellite photography, the Space Operations Support Division of the Weather Bureau provided the forecast that resulted in shifting the landing point of Apollo 11 about 200 miles to avoid thunderstorms—which were later in fact observed. On a much broader scale, weather data are provided to pilots, controllers, and air service operators almost around the clock; for example, terminal forecasts are made for 425 United States airports every 6 hours. Shipping managers, coastal industry workers, fishermen, farmers, foresters, ranchers, builders, ski resort operators, and those of a myriad other professions and trades are ever more likely to count on reliable and continuous weather information, an increasing amount of which stems directly from space observations. The value of such observations cannot be assessed strictly in terms of dollar savings, since many of the benefits are more readily expressed as a function of increased comfort, greater safety, more enjoyment, or greater efficiency of operation. But even if only financial savings were considered, they would amount to billions of dollars each year.

8
COMMUNICATIONS ON EARTH VIA SPACE

WHEN ABRAHAM LINCOLN WAS ASSASSInated in 1865, the news took 12 days to reach Europe. Ninety-eight years later when John F. Kennedy was assassinated, the world knew of it immediately. The news traveled to the farthest corners of the Earth at the speed of light. Television viewers the world over, who saw his funeral as it took place, saw it only because of communications satellites. Unquestionably the greatest economic and cultural impact, and perhaps the only viable industry, to result from the opening of the "'vertical frontier"–as Kennedy called space–is in the field of communications.

Satellite communications is the one commercially exploitable area opened by space research that early attracted the capital of the entrepreneur in sums comparable to those spent by the federal government. Indeed, this mode of communications developed so swiftly that some 600 million people, almost 17 per cent of the Earth's population, observed the first landing of men on the Moon as it actually occurred. And this event took place only 12 years after the first, crude artificial satellite of Earth had been launched.

The alacrity with which private capital rose to support the communications satellite stands starkly in contrast to the attitude expressed during a similar "quantum leap" in communications technology during the late nineteenth century. Sir William Preece, chief engineer of the British Government Post Office, expressed his view of Alexander Graham Bell's recent invention and its potential with: "'The Americans have need of the telephone—but we do not. We have plenty of messenger boys."

The number of international phone calls is estimated to rise at an annual rate of 20 per cent. In 1930, only 32,000 were made. By 1960, however, this number had grown to 3½ million. The projected number for 1980 is variously estimated between 40 million and 100 million. As 1970 dawned, one of every two overseas telephone calls was via satellite.

In general, technological progress and communications go hand in hand. Thus the communications satellite seems assured of a continuing place in the world economy because of its efficiency and its adaptability to growing needs for more channels with wider bandwidths. Today a phone call from Rio de Janeiro to Caracas must be routed through New York—unless it goes by satellite. Even though such messages travel almost instantaneously, there is a basic inefficiency in the old system.

The first idea for a communications satellite per se seems clearly to belong to Arthur C. Clarke, the well-known writer of science fiction who was also instrumental in the development of radar during World War II. In the February 1945 issue of *Wireless World*, he suggested that artificial satellites of the Earth "could give television and microwave coverage to the entire planet." In the October issue of the same magazine, he elaborated his idea with an article entitled "Extraterrestrial Relays." In it he proposed the system of communications satellites that is now a reality provided by the Intelsat Consortium. Curiously, however, Clarke saw the satellites as manned. The idea of the unmanned communications satellite apparently was conceived by Eric Burgess, an English engineer, who suggested it in the November 1959 issue of *Aeronautics*.

As farsighted as Clarke's proposal was, he overlooked the fact that the Earth already had a communications satellite in orbit awaiting man's exploitation. In fact, it had been in orbit for some 5 billion years or more. As early as 11 January 1946, the U.S. Army Signal Corps bounced radar signals off the Moon during Project Diana. On

29 November 1959, voice transmissions were relayed from Holmdel, New Jersey, to Goldstone, California, via this same natural communications satellite. The Moon was also used as a reflector of radiowaves by the U.S. Army in the 1950s when existing channels between the United States mainland and Hawaii failed because of atmospheric disturbances.

Despite the a priori claims of the Moon, the world's first true communications satellite is generally conceded to be the result of Project Score, a joint venture of the U.S. Department of Defense's Advanced Research Projects Agency and the Convair Division of General Dynamics Corporation in 1958. In truth, it was more of a propaganda stunt to save at least some face for the United States following the successful launchings of the Russian Sputniks 1, 2, and 3 and the failures of various United States–produced Vanguards and Pioneers. On 18 December 1958, an Atlas intercontinental ballistic missile was launched into Earth orbit. Attached to its empty hull was a 15-pound pod that contained two radio receivers and transmitters and a tape recorder, powered by batteries.

The tape recorder broadcast a cheery Christmas message from President Eisenhower to the peoples of the Earth. However, the Score vehicle was also used to transmit messages between Texas, Arizona, and Georgia for 12 days before its batteries ran out. In a technical sense, it was a "delayed repeater" satellite: one that could receive a message upon passing over a particular spot on the Earth, store it on tape, and then "dump," or transmit it on command, over another spot on the Earth. Since the burned-out Atlas stage weighed some 8,700 pounds, the Score craft remains today the heaviest communications satellite ever orbited by either the United States or the Soviet Union.

THE ROLE OF THE COMMUNICATIONS SATELLITE

The satellite overcomes the problems involved in spanning oceans and continents with submarine cables, land lines, and microwave radio relay stations for the long-distance transmission of radio, telephone, and television signals. Since the mi-

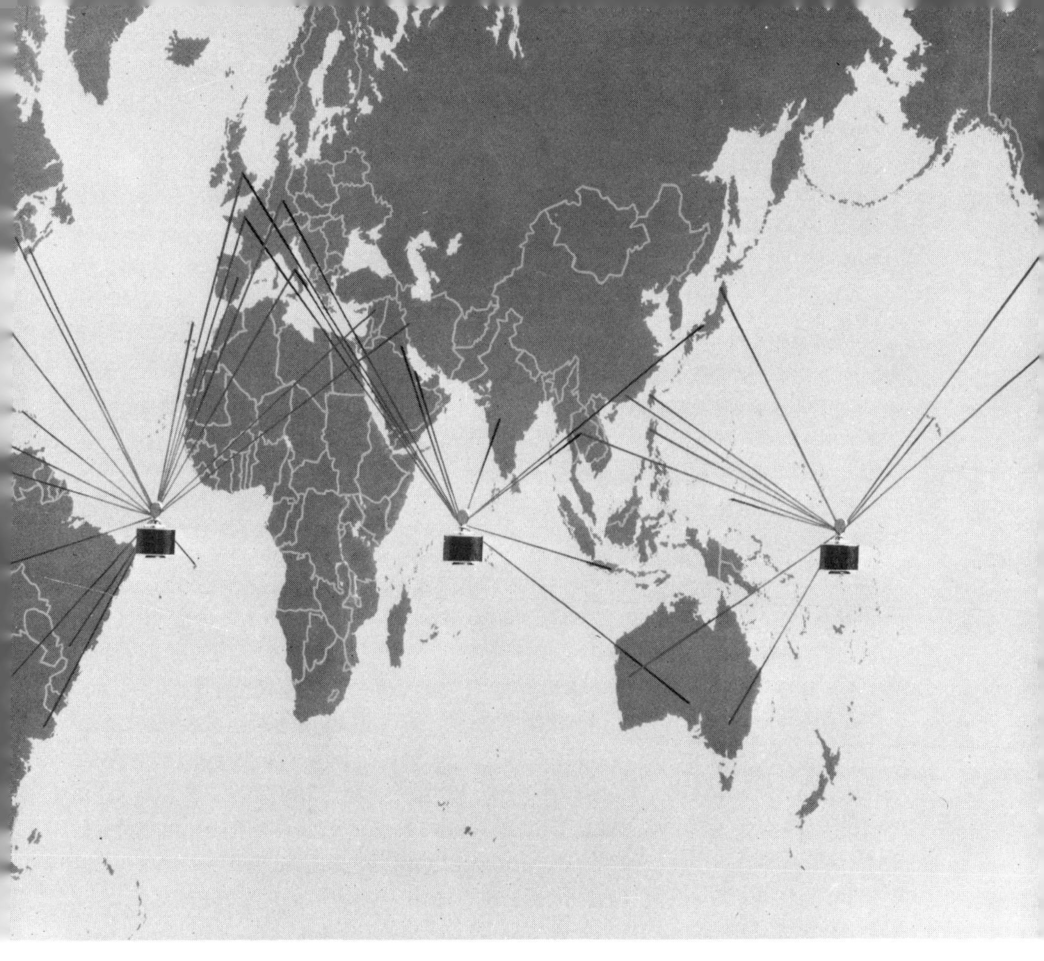

Three communications satellites placed in synchronous orbit above the Atlantic, Pacific, and Indian oceans provide a network that effectively covers the entire globe, except for small portions at the north and south poles. (COMSAT)

crowave travels in a straight line, relay stations to receive, amplify, and retransmit signals must be spaced every 35 miles or so because of the Earth's curvature. The cost of providing enough stations to encircle the world completely would be prohibitive. (A *single* relay in the Atlantic Ocean would have to be over 400 miles high to be "seen" from the coasts of America and Europe.) Submarine cables, likewise, are extremely expensive, and they tend to reach the limit of their capacity immediately after installation. However, they do have the advantage of not taking up space in the already crowded radio-frequency spectrum allotted to satellites.

A satellite at a distance of 22,300 miles above the Earth (at the equator) is in a synchronous orbit—it appears to remain stationary over a particular spot on the Earth. It becomes a relay station that can be seen by any two points on a hemisphere. Three satellites spaced at 120-degree intervals about the equator effectively cover the entire globe, except for very small portions at the north and south poles. Thus, placing a satellite each above the Atlantic, Pacific, and Indian oceans provides a worldwide network. Unlike Earthbound communications modes that operate between two points only, satellites link *all* stations in the area they cover.

Factors other than those dictated by the laws of physics also promoted the development of the communications satellite fairly early in the space era. Typical of these was the concern for the reliability of intercontinental radio communications in the interests of national defense.

Conventional radio contact over long distances is unpredictable because it depends upon the ionosphere, which reflects high-frequency radio waves from the surface, bending them back to the Earth. The ionosphere is subject to a variety of disturbances that play havoc with radio waves, causing static and fadeout of reception. Sunspots, which occur at unpredictable times, are particularly disruptive of the ionosphere. Likewise, atomic bomb detonations in the ionosphere can affect radio communications. Therefore the need for maintaining contact among armed forces and the Pentagon during a nuclear war also spurred research in the field of satellite communications.

With the advent of the communications satellite, the use of superhigh-frequency microwaves became possible. Unlike the high-frequency radio waves that are reflected from the ionosphere, the superhigh-frequency waves pass through it and out into space. Their use with a satellite provides for broad-band circuits which transmit much more data, and they open up new frequencies for use in communications. This broad-band capacity and the synchronous altitude are the unique features of the communications satellite that make live transoceanic television possible.

The extremely high overall reliability of the communications satellite system as compared with the cable, land line, and microwave relay network also argues well for its continued expansion. Each year Intelsat realizes quite considerable profit from the use of its satellites by communications companies when their cable or land line service is interrupted. In 1968, for example, there were 24 occasions on which such interruptions saw carriers leasing satellite circuits in order to maintain service.

Satellites do go "off the air," however. They lose service 12 times a year, for a period of about 8 minutes a month. These occur when the satellite is directly between the Sun and a particular Earth station; the high noise level from the Sun simply drowns out signals from the satellite. But these particular 8-minute periods are known in advance, and alternate routings are made.

Communications satellites follow the classic law of supply and demand in economics of operation. As more satellites become available, the service charge decreases accordingly. George P. Sampson, vice president for operations of Comsat, in addressing the National Cooperative Telephone Association on 6 February 1970 in New Orleans stated:

> Considering economics as it relates to the cost of the satellites, our first generation satellite, known as the Early Bird, indicated a cost per circuit-year of some $15,300. This was the pioneering commercial satellite launched in 1965. The second generation launched in late 1966 reduced the cost per circuit-year to about half. The third generation, the ones we are now using, launched during the course of the last fourteen months, has further reduced the circuit cost per year by a factor of four. . . . Also, regarding economics, once a nation has paid for its Earth station—and this price of admission to the system is going down each day—that nation can establish direct links with other nations at sharply declining costs. The cost for an additional circuit or path is hard to find.

THE DEVELOPMENT OF THE COMMUNICATIONS SATELLITE SYSTEM

The first scientific experiments with satellite communications involved passive systems—those that merely reflected unamplified signals. Echo 1, launched on 12 August 1960, was a 100-foot-diameter balloon made of aluminized mylar plastic only 0.0005 inch thick. It orbited the Earth at an altitude between 945 and 1,049 miles, providing a reflector for radio and television signals beamed at it. For several years the satellite found use in such a ca-

Scientific experimentation with satellite communications began with passive systems such as that employing Echo 1, which merely reflected signals instead of receiving, amplifying, and retransmitting them. A 100-foot-diameter balloon, Echo 1 was launched in 1960 and remained in orbit until 1968. (NASA)

pacity; but it also provided another dividend. Geodesists around the world used it to plot geographical locations with far greater accuracy than had been thitherto possible using Earthbound geodetic methods. Gradually micrometeoroids began to puncture the balloon, and the pressure of the sunlight pushed it into lower orbits. Wrinkled and sagging, it reentered the Earth's atmosphere and blazed to death as a "shooting star" in 1968. Echo 2, an improved model, was orbited in 1964 and signals were relayed by it from the United States to the

Soviet Union. The entire Echo program cost only $17 million, including launch vehicles.

Between 1960 and 1964, several experiments with active satellites (those that receive, amplify, and retransmit signals) were made in the United States by the Department of Defense, American Telephone & Telegraph Company, and the National Aeronautics and Space Administration. In addition to Score, the Department of Defense also developed the Courier. A delayed repeater satellite, Courier weighed 500 pounds and incorporated 19,200 solar cells on its spherical surface to supply its internal electrical energy. It could receive, store, or transmit some 68,000 words per minute. Launch took place on 4 October 1960 with a Thor-Able-Star rocket from Cape Canaveral (now Cape Kennedy), Florida.

The potential of the communications satellite soon became obvious to American industry as a result of the experiments made by the relatively primitive Echo 1 and the more advanced Courier. On 19 January 1961, American Telephone & Telegraph Company was authorized by the Federal Communications Commission to establish an experimental communications satellite link across the Atlantic Ocean at its own expense. The satellite was designed and built by Bell Telephone Laboratories and launched by NASA, which was reimbursed by AT&T for the booster and associated launching costs.

The result of this effort was Telstar 1, which far exceeded its designers' fondest hopes. Launched on 10 July 1962 by a Thor-Delta rocket, it relayed live television pictures from the United States to France on the same day, and for six months was used for engineering research and public demonstrations. Communications engineers learned much from Telstar, including the fact that the deadly radiation of space can put solar cells of such satellites out of commission unless they are properly shielded. Telstar 2 was launched on 7 May 1963, and proved a complete success. Transmissions between stations at Andover, Maine; Goonhilly Downs, England; and Pleumeur-Bodou, France, showed that such satellites could be tracked by large antennas with very narrow beam widths (0.1 degree). Telstar 2 mysteriously went off the air from 17 July to 12 August, but then corrected its malfunction and operated until May 1965 before going dead. Still circling silently around the Earth, it will reenter the atmosphere, according to estimates, as a shooting star in the year 11963.

As Telstar was on the drawing boards, NASA also advanced the state-of-the-art of communications satellites by sponsoring Relay, which was developed under contract by Radio Corporation of America. The objectives of Relay were about the same as those of Telstar, but for reliability it had two transmitters in case one failed. These were more powerful than those of Telstar, which meant that a less powerful ground station could be used. Launch took place on 13 December 1962. Among the various accomplishments of this pioneering satellite were the live televised transmission to Europe of the unveiling of the *Mona Lisa* on its first visit to the United States; the first transmission of telephone conversations via satellite between the United States and Italy, South America, and West Germany; and the first television transmission via satellite between the United States and Japan. Relay 1 also showed the feasibility of setting type in Britain using computerized data generated in Chicago, and accomplished the first transmission of brain waves from a patient in England for analysis by a doctor in the United States. Relay 2, orbited on 21 January 1964, embodied improvements in technology; it transmitted until September 1965. The Relay program, including launch vehicles, cost $37 million.

The satellites discussed so far were placed into medium-altitude orbits. Since Clarke's 1945 article, engineers had known that the optimum altitude for communications satellites is some 22,300 miles above the equator. They could not take advantage of synchronous orbits, however, until rocket boosters powerful enough to deliver heavy satellites to such altitudes became available.

The first successful communications satellite to be placed in synchronous orbit was Syncom, launched on 23 July 1963. A joint program of NASA (which spent $26 million on it, including boosters) and the Department of Defense, Syncom firmly proved the feasibility of the 24-hour communications satellite. During its developmental employment, various kinds of signals were sent to and from the craft via ground stations and ships at sea. The performance of the Syncoms convinced scientist and entrepreneur alike that the communications satellite was practical and potentially profitable. The technology developed in the Syncom program was especially applicable to the satellites used by the world's first commercial communications-satellite network.

COMSAT, INTELSAT, AND INTERSPUTNIK

The remarkable series of successes by the early communications satellites, such as Score, Echo, and Courier, was clear evidence to the public that here was an aspect of space technology that could benefit all and possibly pay its own way. Congress passed Public Law 87-624, the Communications Satellite Act of 1962, on 27 August. It was signed into law only four days later by President John F. Kennedy, who said prophetically upon the occasion: "The ultimate result will be to encourage and facilitate world trade, education, entertainment, and many kinds of professional, political and personal discourses which are essential to healthy human relationships and international understanding." The act provided for the establishment of a privately owned corporation to serve as the only United States entity in international communications satellite agreements.

The articles of incorporation for Communications Satellite Corporation (Comsat) were signed on 29 January 1963 and approved by the President three days later. The certificate of incorporation was issued in Washington on 1 February, and the first officers were elected on 10 March. On 24 June 1964, an issue of $200 million in stock was offered and *oversubscribed*. Half of the issue, as specified by law, was offered to the public and half to American common communications carrier companies. Each segment bought 10 million shares. By 1970, 132,000 individuals owned 62 per cent of the stock, and 117 common carriers owned 38 per cent.

In its international role, Comsat became the manager of operations for the space segment of the International Telecommunications Satellite Consortium (Intelsat). This role included the design, development, operation, and maintenance of the satellite system. Intelsat is a multinational organization consisting of members of the International Telecommunications Union, formed by two international agreements opened for signature in Washington on 24 August 1964. Eleven nations signed on the first day. By early 1970 the number of signatories had risen to 74, representing over 96 per cent of the world's telecommunications traffic.

Comsat's first major venture in Intelsat's space segment was to contract for the development and launch of a communications satellite. The result was the Early Bird, built by Hughes Aircraft Company and drawing heavily on the technology of the Syncom. On 6 April 1965, Early Bird was launched from Cape Kennedy by a Thor-Delta booster, purchased by Comsat from NASA for some $3 million. The satellite, intended to last 18 months, astounded both its designers and its owners by operating far beyond this design requirement.

Early Bird's 240 two-way voice circuits were equal to the capacity of two existing transatlantic submarine cables, and it provided an increase of more than 80 per cent in the capacity for high-quality voice communication across the Atlantic Ocean. For 42 months Early Bird remained in commercial service, often acting as a standby satellite for later, more sophisticated models and taking over when necessary. It was phased out of service on 18 January 1969, with a record of 100 per cent reliability over the period of its operation.

In 1967 three Intelsat II satellites (built by Hughes Aircraft Company) were placed in orbit—two over the Pacific Ocean and one over the Atlantic. Intelsat II model F-2, launched on 11 January, provided the first full-time service over the Pacific, and communications began immediately back and forth between Japan and the United States. The Intelsat II satellites each had, like Early Bird, 240 two-way voice circuits, but they radiated twice the power and their antennas covered twice the geographical area. Full global coverage for the Intelsat system was not achieved until 1969, when three Intelsat III satellites (built by TRW, Inc.) joined the Intelsat II models already in orbit. Not only were these heavier, but each provided 1,200, rather than 240, two-way voice circuits. They also had almost 10 times the radiating power of the Intelsat II models.

With the successful launching of the first Intelsat IV satellite on 25 January 1971, the technology of communications by satellite took a giant leap forward. A product of Hughes Aircraft Company, it also incorporated significant contributions from aerospace companies in the United Kingdom, West Germany, Switzerland, Belgium, Japan, Italy, Sweden, Spain, France, and Canada.

In describing the significance of the new satellite, the Intelsat IV program manager for Hughes said: "One of these spacecraft will have a capacity far exceeding the total of all the world's submarine cables. One Intelsat IV will have an average capacity of 6,000 two-way

The new generation of Intelsat IV satellites have steerable, dish-shaped antennas that focus power in beams directed at heavily populated areas, where communications needs are greatest. (COMSAT)

telephone circuits or 12 simultaneous color TV programs, or tens of thousands of teletype circuits, or combinations of these."

This versatility is possible because the Intelsat IV has two steerable antennas that focus power in beams, which can be directed at particular areas of the Earth where communications needs are greatest. The effective radiated power is between 8 and 84 times greater than that of the Intelsat III series.

Similar advances in technology, with concomitant reductions in costs, are evident in the ground segment of Intelsat. The ground station at Andover, Maine—one of the first built by American Telephone & Telegraph Company for use with its Telstar, and later sold to Comsat—cost some $15 million in 1962 and required about 50 people to operate it. By 1970, a typical ground station cost only $3 million to $5 million and needed only 12 to 14 operating personnel.

The Soviet Union in August 1969 proposed to the UN Committee on the Peaceful Uses of Outer Space a competing international communications satellite network. The new system, Intersputnik, would be another joint venture, with Bulgaria, Cuba, Czechoslovakia, Hungary, Poland, Romania, Mongolia, and perhaps East Germany as the initial members.

DOMESTIC COMMUNICATIONS SATELLITES

In addition to Intelsat, several nations have domestic satellite systems planned or under development. The earliest such network deployed was the Orbita television system of the Soviet Union, which became operational in November 1967. By 1970 Orbita was covering about 65 per cent of Soviet viewers, and it was estimated that within five years some 85 per cent would be reached. The system apparently is still not without some flaws. In a rare admission of faulty technology, *Pravda* on 8 February 1970 complained of the unpredictable service offered by Orbita. The journal pointed out that on 10 December 1969, for example, a program scheduled to begin in Chita at 5 P.M. was more than an hour late. After 15 minutes of news there was to have been a children's program, but a travelogue came on instead. And the next two programs were unscheduled, with the transmission ending 10 minutes early.

France, West Germany, and Belgium are combining resources to develop the Symphonie satellite, which will give them valuable experience in the technology of communications satellites. Indeed, beyond Symphonie, France already has plans for a wholly French system. In Project Dioscures, she hopes to launch five domestic satellites into synchronous orbit by 1972, stationing them over the Atlantic and Pacific oceans. By 1977 she hopes further to have five second-generation satellites in a global network. Dioscures will provide telecommunications relay as well as air traffic control and navigation aid when fully developed. The initial system is estimated by the French to cost some $156 million. Its realization has been represented by France as a national space goal, much as was Project Apollo by the United States.

Canada plans to have Anik 1, the first of her three domestic satellites, in orbit by 1972. The second satellite is scheduled to be operational by mid-1973, with commercial traffic beginning early in 1973. Japan has similar satellites under consideration, and countries such as India, Pakistan, Brazil, Australia, and Indonesia are also considering them.

The reasons why these countries desire their own communications satellites, and are willing to invest large sums in them, are varied.

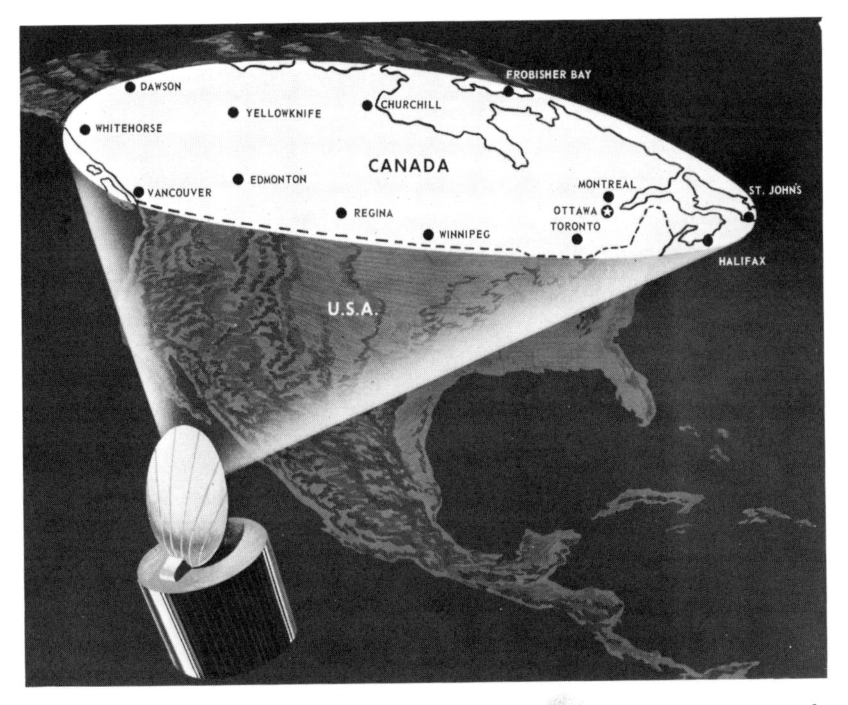

Canada plans to establish a communications satellite system for purely domestic use. A satellite in synchronous orbit above the equator could provide telephone and color television service from coast to coast. (HUGHES AIRCRAFT CO.)

Canada, for example, explained her reasons for developing a domestic system in a government "white paper" issued in 1967. The government of Canada feels that the only economical way to provide telecommunications services to her population is via satellite. The costs would simply be prohibitive if the nation tried to develop a network of land lines and microwave links to tie her widespread population centers together. The satellite network also will assist in the opening of vast areas of uninhabited land to development and exploitation, releasing more funds for road and airfield construction and waterways improvement. Telesat, the Canadian equivalent of Comsat, is expected to spend some $90 million on a system with two satellites in synchronous orbit, one spare stored on the Earth, and 37 ground stations.

Pakistan finds herself a nation of two halves, with several thousands of miles of less-than-friendly India between. The construction of conventional land lines and microwave links to tie the two halves together is clearly more than a problem of money. The existing radio link between East and West Pakistan is of course subject to disruption because of fluctuations in the upper atmosphere.

Indonesia's population is scattered among some 3,000 islands; the costs of submarine cables and microwave links are prohibitive. Australia and Brazil face a problem akin to that of Canada—vast hinterlands and a widely spread population. Other countries, such as Chile and Colombia, could also well utilize a communications satellite, perhaps on a regional basis with other South American countries. These nations are divided by the Andes Mountains, a formidable barrier to cross with land lines and microwave links.

Private industry in Italy began in 1968 to fund the development of the 800-pound Sirio satellite for telecommunications experiments, looking forward to a launch in 1972. While not an operational communications satellite per se, Sirio will provide much-needed experience for the Italian aerospace industry and valuable scientific data for the design of Italian satellites of the future.

The one factor not clear in the plans of France, Canada, and West Germany is the means for orbiting these satellites. Launching vehicles currently are available only from the United States and Soviet Union—and perhaps in the near future from the European Launcher Development Organization, should it survive. Latecoming countries may find the United States unwilling to aid in orbiting communications satellites that will compete with Intelsat or with future American systems. Nor is it likely that the Soviet Union will provide launching vehicles for systems that could compete with its Orbita or Intersputnik system.

As early as 1966, Comsat began studies of potential users of domestic satellites in the United States. These included discussions with communications common carriers, broadcasters, news wire services, newspapers, computer services companies, Western Union, airlines, and community-antenna television companies. Such satellites seem best suited for distributive service—where a satellite in orbit would receive, say, a television program and send it to an almost limitless number of relatively inexpensive receive-only ground stations. However, the studies also looked into other possibilities, such as multipurpose satellites that could serve the needs of several different types of customer.

The service costs for such a system would be low in comparison with the costs of launching a single-purpose satellite for each customer.

Another potential customer for the domestic satellite in the United States is the Postal Service, which is presently considering the communications satellite as a mode of transmission in its proposed Mailgram service. In such use, messages would be sent from one post office to another via satellite and delivered in the regular mail. Results from a two-year experiment begun in 1970, with the participation of Western Union Telegraph Company, may help to determine the future of the proposal. In the experiment, Telex machines belonging to the company were placed in 6,100 businesses and 113 post offices across the country. Messages were sent by conventional land lines rather than satellite; otherwise, the procedure was the same as for the projected system.

A similar scheme has been proposed by General Electric Company. Its Telemail service would provide instantaneous communication via satellite among businesses equipped with special send-and-receive machines. The company, perhaps overoptimistically, estimated that the cost of a 600-word letter would gradually drop to "about 10 cents." Yet another suggested use for domestic satellites comes from the International Board of Trade, in Los Angeles, which has petitioned the Federal Communications Commission for a novel application. This nonprofit corporation is a computerized commodity exchange dealing primarily in metals. It would like to launch two communications satellites for transmitting computerized information between seven ground stations. The total investment proposed for the system is some $55.1 million.

In January 1970, the three major broadcasting networks in the United States began a study to determine whether it would be cheaper to acquire and operate their own broadcasting-only satellites or to lease service from a domestic satellite owned by Comsat or American Telephone & Telegraph Company, for example. Similarly, AT&T announced plans to orbit its own satellites to supplement its network of land lines, cables, and microwave links. Intended primarily as a backup for the Earth system, the satellites could also be used to adjust peak traffic loads and for lease to other common carriers and the radio and television industry. Hughes Aircraft Company, which has much experience in designing and constructing communications satellites, was another organization that began to consider whether it might not be profitable to own its own satellites and sell their services.

This rush into synchronous orbit poses a technical problem that can grow to significant dimensions in a relatively few years. Space may be infinite; but the amount available for communications satellites unquestionably is finite. The specter of overcrowding the synchronous orbits appears. This is one of the major reasons that Canada is hastening development of her domestic satellite. Canadian scientists point out that there are only six or seven "parking" locations in a synchronous orbit suitable for a Canadian system—and these same positions are also suitable for countries in South America as well as for the United States. Another serious effect of overcrowding is that satellites less than 2 degrees apart in orbit can present problems of radiofrequency interference.

MILITARY COMMUNICATIONS SATELLITES

At first glance there would appear to be no significant difference between military and civil satellites. However, the military communications satellite system must possess a flexibility not required by its civilian counterpart. It must be capable of rapid expansion of service within a localized area of operations, and must have a high degree of reliability. The ground segment must be mobile and able to operate in a variety of climatic environments. Most importantly, the system must be secure from enemy monitoring or jamming.

Military communications satellites are generally classified as either strategic or tactical, following classical military terminology for missions and roles of an armed force. The strategic satellite system has satellites tied in with a group of fixed Earth stations, so that communications can be established over any path by which two or more stations are mutually "visible" with respect to a satellite. The tactical system, on the other hand, consists of satellites that permit communications between a variety of Earth stations which are for the most part mobile. These may be in aircraft, ships, automotive vehicles, or even on the backs of soldiers. The operational area, also, is more localized than that of the strategic satellite network. In addition, strategic systems usually are "on the air" full time, while tactical systems operate sporadically.

The need for communications security precludes the use of commercial satellites by the U.S. Department of Defense. Thus, provision for a separate satellite network for national defense was early recognized as a necessity, and it is the one area of the defense budget that seems relatively immune from the budgetary axe wielded by Congress.

The first such network was the Initial Defense Satellite Communications System (IDSCS). Realizing the need for high reliability, the Department of Defense, in conjunction with Philco-Ford Corporation, developed an amazing satellite with no moving parts. The satellite weighed only 100 pounds and was but 36 inches in diameter. Power was supplied by 8,000 solar cells mounted on the exterior surface. The IDSCS was developed between 1966 and 1968 by distribut-

The 7-foot antenna of this mobile terminal for military satellite communications can be stowed out of sight when the jeep is in transit. The trailer carries the electric generator and its fuel. (U.S. ARMY)

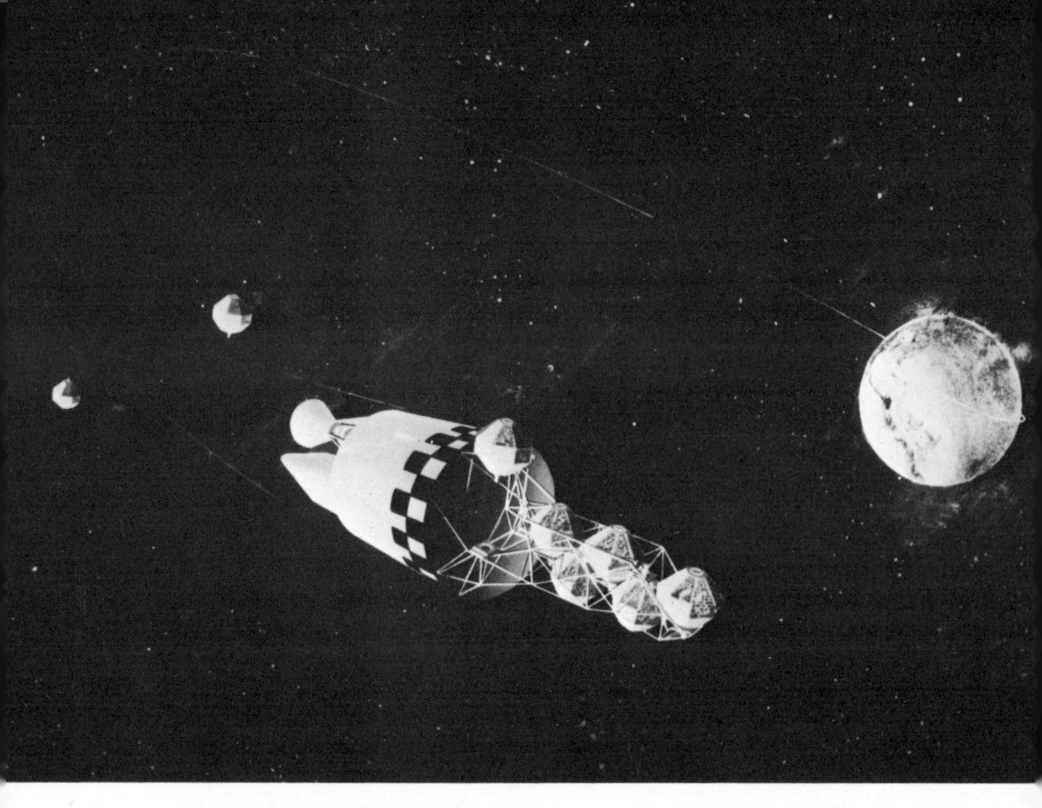

The small (36-inch-diameter) satellites of the Initial Defense Satellite Communications System are launched in batches into near-synchronous orbit, causing them to drift in relation to ground stations. If a particular satellite should fail, the system will continue operating because another satellite is always in view. (U.S. AIR FORCE)

ing 26 of these satellites in random orbits that were slightly less than synchronous. Each satellite drifts some 30 degrees per day relative to a point over the equator. Each, then, stays in view of an Earth station for about 4½ days before drifting out of "sight." If a particular satellite should malfunction, the system is not interrupted because another is always in view.

The IDSCS satellites were designed with an expected lifetime of 3 years, but by 1970 it was apparent that they would "live" much longer. In fact, each had a built-in device for automatically shutting the satellite off after 6 years. By 1967 the system was regularly transmitting high-speed digital data from South Vietnam to Washington. High-quality reconnaissance photographs made in battle zones were available to analysts in the Pentagon, via IDSCS, within minutes after being processed.

The tactical communications satellite, unlike the strategic, must be very large and have much more power. Because the strategic system utilizes very large, immobile, high-power Earth stations, its satellites can be small and simple. The tactical satellite, however, operates in a network of mobile, low-power Earth stations and consequently must be larger and radiate considerably more power. The first operational tactical satellite, Tacsat 1, was launched into an Earth-synchronous orbit from Cape Kennedy on 9 February 1969. Weighing 1,600 pounds, it was 25 feet in length and 9 feet in diameter. Tacsat 1 had three antenna systems, each of which could be pointed to a different area of the Earth. It also had an effective radiated power of 6 kilowatts as compared with the 3 watts of the IDSCS satellites. The communications capacity of Tacsat 1 was comparable to that of 10,000 two-way telephone channels.

Advanced types of military satellites have also been orbited by the United Kingdom and the North Atlantic Treaty Organization. These satellites are in synchronous orbits but have a reaction control system on board that permits them to be moved to different locations around the orbit. They contain batteries powerful enough to insure functioning even in a total eclipse.

The communications satellite has made an important contribution to national defense and will continue to play a leading role in military planning. With weapons systems becoming more sophisticated and with international relations remaining complex and tenuous, the need will inevitably grow for a communications system that has the flexibility to react quickly to crises anywhere in the world. There simply will not be time to lay land lines, microwave links, and submarine cables or to develop other conventional modes of communication. Only the communications satellite can satisfy this important need.

THE POTENTIAL FOR COMMUNICATIONS SATELLITES

As the 1960s ended, a wide variety of potential uses were being discussed for the communications satellites of the following decade. Comsat studies reveal that the satellite of the future may be able to do such things as control commercial air traffic, link computers into worldwide networks that will be 600 times faster than present teletype circuits, provide a worldwide medical diagnostic service for doctors, and transmit facsimile copies of entire books or newspapers from one nation to another in minutes. It should be possible also to provide direct-dialing telephone service between countries.

NASA's applied technology satellites have been used for testing the transmission of television and radio programs within the United States; for reliable ship-to-shore communications over long distances; for communicating with aircraft in transoceanic flight; and for relay of information from remote, unmanned scientific instruments and buoys.

Undoubtedly the next generation of communications satellites will see a direct-broadcast model in which radio and television programs can be transmitted from the satellite either directly to the home set antenna or to a central antenna and thence by cable to the home. Paradoxically, the direct-broadcast satellite seems to have a greater potential in the underdeveloped emerging, hence "poor," country than in the nation with highly developed communications systems. The cost of developing and launching such a satellite system for a country like India would be only a fraction of what would be required to set up a conventional telephone, radio, and television network like that in the United States today. In the United States, on the other hand, one must consider the economic consequences of such a departure on the existing system: what would happen to all the television and radio stations now in operation?

Perhaps the most often-mentioned use of the direct-broadcast satellite is for mass education. By the beginning of 1970, India, Brazil, and Indonesia were planning for such satellites. Brazil, for example, submitted a plan to NASA for the inclusion of an educational television experiment aboard one of its applications technology satellites to be launched in the mid-1970s. The satellite would relay programs, sent up to it from a station at the Federal University in Natal, to some 500 receive-only antennas set up at schools throughout the state of Rio Grande do Norte.

India is involved in a similar cooperative program with the United States. A special television experiment aboard ATS-F, a satellite to be launched by NASA in 1973, will beam educational programs throughout this large nation. The system will also permit the Indian government to propagandize for birth control, and to televise information on modern methods of agriculture to marginally productive or inefficiently worked areas of the country.

The plan also will boost the industrial potential of the country, because the television sets needed for the ambitious program will be built in India—hopefully at an ultimate rate of 10,000 a year. Initially, the experiment will involve placing some 5,000 community sets in selected villages. Subsequently it will be expanded to include a set in each of the country's 560,000-odd villages, even those without electricity—which will be supplied for the purpose by a man on a bicycle-powered generator! (It is estimated that only 2 per cent of India's population will ever be able to afford private television sets.)

In the case of India, and several other countries as well, the direct-broadcast satellite can be a powerful tool for helping to unite a people linguistically. In India alone, for example, there were, as of 1968, 800 recognized languages or dialects spoken by some 523,893,000 people.

While much speculation and planning have been done in the area of educational television via satellite, the use of radio in a similar role has been largely overlooked. A study made by H. J. Skornia, an electrical engineer who reported his findings during the Institute of Electrical and Electronic Engineers' International Conference on Communications in 1968, shows that radio instructs as effectively as television and costs only 5 to 10 per cent as much per channel. While there may be some argument as to the validity of the assumption of its effectiveness, there can be no doubt that radio would be much cheaper. A radio educational system based on the direct-broadcast satellite could well bring mass education to nations without the resources for a television system.

While satellites will continue to be an invaluable means of communication, they will not replace the submarine cable. To meet the continuing demand for more channels and circuits, reliance on these cables in setting up intercontinental links will be intensified in the future. During the first five years of commercial communications satellite service, the number of cables trebled. By the end of 1970, some 94 cable systems were in operation over more than 85,000 nautical miles. From the economic viewpoint, we can say that the satellite offers advantages in wideband transmission such as television, and in speech circuits where the traffic is relatively light. It is particularly applicable for diverse, low-density, long-haul routes. The cable, on the other hand, is more economical for heavy trunk traffic routes.

Let us look to an undetermined date in the future:

> [Ultimately we shall] be able to visit all museums, read any book in any library, attend all first nights, call up the knowledge of the ages, stored in the memory circuits of giant computers. Communications satellites will end ages of isolation, making us all members of a single family, teaching us to read and speak, however imperfectly, a single language. Thanks to a few tons of electronic gear 20,000 miles above the equator, ours will be the last century of the savage—for all mankind, the Stone Age will be over.

Is this a view of some "brave new world" as forecast by Marshall McLuhan? No—it is yet another speculation by Arthur Clarke, who suggested the idea of communications satellites to an incredulous public back in 1945.

9

RESEARCH IN SPACE

PRESIDENT JOHN F. KENNEDY COMPARED space to the ocean when he decided in 1961 to accelerate the Earth-orbital, lunar, and planetary exploration programs. He emphasized, in simple terms, that "we must" sail on this new ocean of space. No profound intellectual, moral, or practical reasons were given, though political and military factors obviously were involved. The challenge of achievement, of reaching out with a newfound ability, of accomplishing that which was within our power, was sufficient reason for the young President to set the nation's course toward the stars. The exploration of the ocean of space was, in a real sense, merely the next logical step in man's age-old quest to understand the nature of the universe—a step that man had to take unless he chose to limit his curiosity and set arbitrary boundaries to his intellectual horizon.

By launching telescopes and other instruments into space, scientists have gained an entirely new view of the universe. From the ground, the view is hazy at best. Clouds block out images; the motion of gases in the atmosphere blurs them; the ionosphere prevents ultralong energy waves from reaching earthbound instruments; the background heat radiation of the Earth obscures measurements in outer space.

Above the atmosphere, the astronomer is in a new realm. The sky

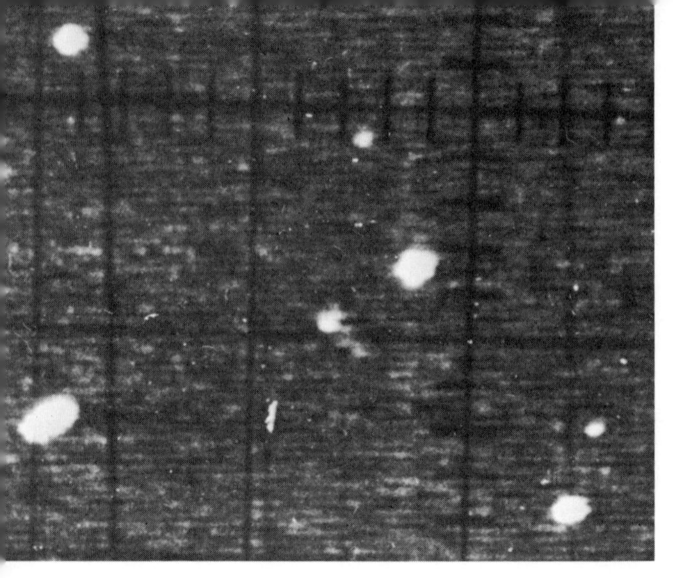

One of the first ultraviolet photographs from the Orbiting Astronomical Observatory, taken in December 1968, shows stars in Constellation Draco. (NASA)

is twice as dark as from the Earth, permitting accurate measurement and analysis of minute light variations and other energy fluctuations. For the first time, he can make observations in the ultraviolet range—especially important in the case of young, very hot stars, which radiate nearly all their energy in this part of the spectrum. Likewise, he can undertake more accurate infrared studies, no longer having to worry about the absorption of these waves by the carbon dioxide and water vapor in the atmosphere. The development of X-ray astronomy, too, has been aided greatly by the use of outer-space instruments. Meanwhile, the magnificent close-up photographs of the Moon taken by the Ranger, Surveyor, and Lunar Orbiter spacecraft, together with the work of the Apollo astronauts, have revolutionized our knowledge of the Earth's nearest neighbor in the solar system. Probes of Venus and Mars have overturned many long-standing conceptions of those planets.

These studies pay no dividends in an immediate, practical sense. No one is able to say just how an improved understanding of the formation of stars, for example, is going to benefit the lives of men on the Earth. This is basic research—by definition, the pursuit of knowledge for its own sake. But, of course, basic research does pay off in the long run by increasing the fund of knowledge at man's disposal. The basic works of Newton, Faraday, and Einstein led to an age of machinery, electricity, and atomic energy. Today, the intellectual stimulation that scientists experience from entering and understanding the space environment has become a major reason for sailing the new ocean of space. And man's penetration of and operation in that environment have necessitated a very broad, interdisciplinary research effort. Fields of investigation range from astronomy and celestial mechanics to analyses of physical, chemical, and biological reactions in the outer space environment.

SOLAR ASTRONOMY

The only star near enough to the Earth for detailed observation is the Sun. Its energy output sustains life on the Earth and is of basic importance to the physics of this planet. For these reasons alone the Sun would be worth studying. Fortunately for astronomers, the Sun also is a typical specimen of the most numerous class of stars, thus giving broader significance to their solar observations.

Scientists have obtained much new and valuable data about the Sun and its environment as a result of the development of space technology. The sources of sunspots and flares, the nature of the Sun's magnetic field, the temperature and density of its gaseous chromosphere and thermosphere—all can be studied much more effectively from above the Earth's atmosphere, which filters solar emissions, than from the face of the Earth itself.

The NASA Orbiting Solar Observatory (OSO) satellite program, active since 1962, has helped to develop important new concepts of the Sun and its activities. Another NASA project, the Apollo Telescope Mount (ATM), which is an element of the Skylab orbital workshop, will permit manned observations from orbit, including long-term measurements on the solar disk in the extreme ultraviolet and X-ray portions of the spectrum and photographing of the solar corona. In order to conduct these observations, the ATM's telescopes will be permanently pointed toward the Sun. The actual images of the Sun, as recorded by the various telescopes, will be visible to the astronauts on two television display monitors, permitting them to pick out specific regions, either targets of opportunity or areas of special

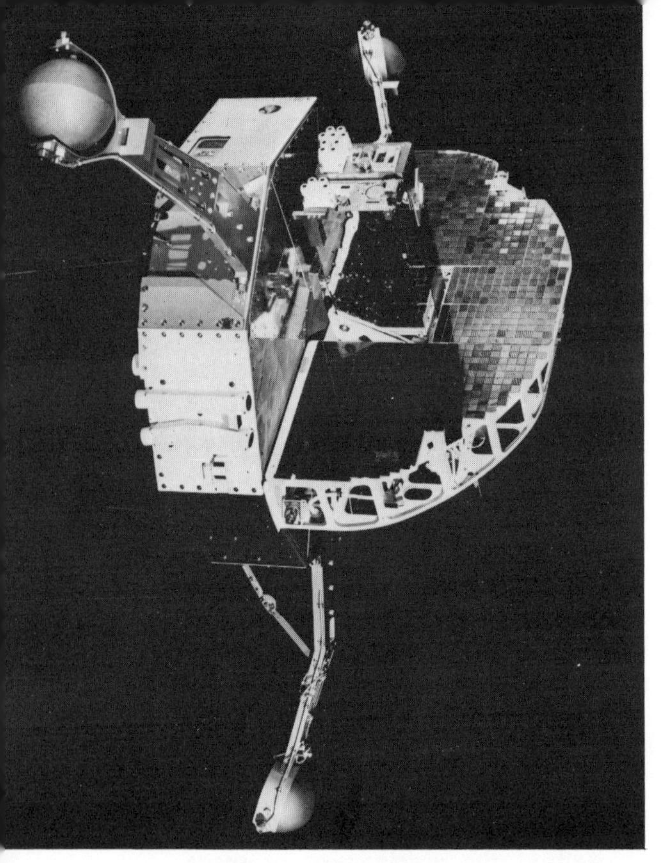

The Orbiting Solar Observatory, designed to study the Sun and its influence upon the Earth. (NASA)

interest, then to examine and traverse them as desired. Data derived from the TV system will also be relayed to Earth control and to astronomers at monitoring stations on the ground.

PROBING THE MOON AND PLANETS

The missions of manned and unmanned spacecraft to the Moon, together with the unmanned probes of Mars and Venus, have paid significant dividends to other projects. For example, the telemetric, optical, and manipulative systems developed for the unmanned Ranger, Surveyor, and Lunar Orbiter spacecraft can be adapted for use here on Earth to record, monitor, and control remote operations, such as the handling of radioactive materials and the investigation of areas that are inaccessible to humans. Likewise, the transportation, navigation, and life support systems developed for the manned Apollo missions to the Moon can be adapted for other spatial and terrestrial programs.

The various missions to the Moon and nearby planets have been primarily beneficial to the pursuits of pure science. Through close-up views and instrument samplings, they have provided more information within a span of a few years than astronomers had gained in all the preceding generations of telescopic observation. The new insights thus obtained into the origin, history, and present condition of the solar system inevitably will lead to a better understanding of planet Earth.

The new findings already have upset many old conceptions. The remarkable composition of the lunar soil, with its high titanium content and unusual distribution of rare elements; the high nutrient "fertilizer" effect of lunar soil on certain plants; the effects on the Moon of the intense bombardment by the solar wind—all these findings were unexpected, and none could have been discovered except by sending vehicles to the Moon. In the near future, exploration of different lunar regions, extended stay-times on the surface, and improved instrumentation will enlarge greatly man's knowledge on the Moon, and perhaps suggest ways in which he can make use of this natural satellite: mining its minerals, or taking advantage of its low gravity and lack of atmosphere for manufacturing operations. Going one step further, it has even been suggested that the establishment of lunar bases might be facilitated by deliberately opening the lunar crust, thus releasing stored vapors and gases to form a low-density atmosphere; this would then be upgraded to an earthlike atmosphere by the introduction of oxygen-producing plants and nitrogen-producing bacteria.

The Mariner "fly-by" probes of Mars have shown that the evolutionary history of that planet may possibly be unique in the solar system. Although generally pocked with craters, the Martian landscape also includes large featureless areas, as smooth as the beds of dry lakes on the Earth. Since it is likely that craters once were distributed evenly over the planet's surface, their absence in these areas implies some later geological event or process. Mariner photographs also reveal some areas with an irregular, jagged topography, similar to that of a landslide area on the Earth. The famous "canals," meanwhile, have proven very elusive; only a limited number of features

Two probes by Mariner spacecraft, numbers 6 and 7, were made of Mars in 1969. This photograph by Mariner 7 was taken from an altitude of about 3,300 miles above the Martian south pole. Clearly visible are three large craters, numerous snowdriftlike formations, and, at upper left, an irregular cloudlike feature. (NASA)

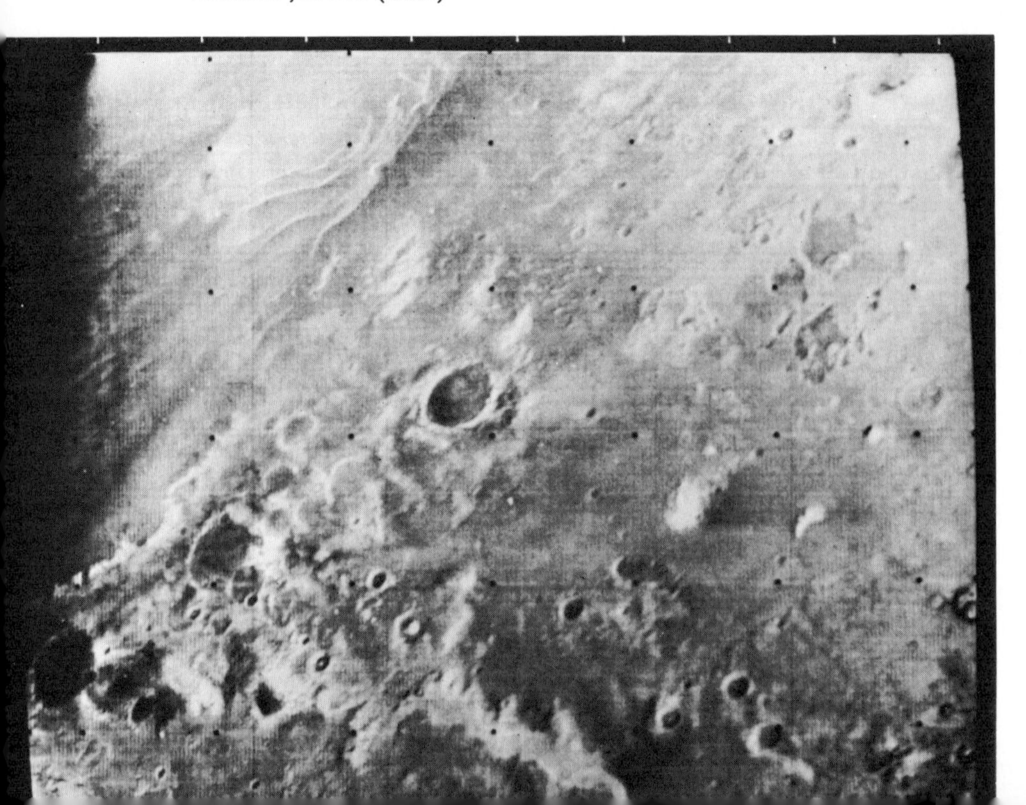

resemble the linear alignments sometimes seen by earthbound astronomers.

Further Mariner probes are planned that will orbit the planet for approximately 90 days. Exploration of Mars is scheduled to continue in the mid-1970s with the Viking program. This will involve orbiting the planet for 140 days, then landing a vehicle on the surface. It is planned that measurements will be made of the composition, temperature, pressure, humidity, and wind speed of the atmosphere; and of the topography and composition of the surface. Also, the type of any organic compounds present, as well as the amount and form of water, will be determined.

It seems unlikely that higher life forms as we know them exist on Mars because of the planet's thin atmosphere, which consists mainly of carbon dioxide; the bombardment of the surface by lethal ultraviolet rays; the apparent lack of atmospheric nitrogen; and the scarcity of water. Still, the potential riches of a whole new world, along with the challenges of the unknown, make it likely that Mars will be an important stepping stone in man's journey to the stars.

Venus, Earth's sister planet, remains shrouded in mystery despite probes of its atmosphere by Russian Venera and American Mariner spacecraft. The planet is enveloped in a cloud of carbon dioxide (more than 90 per cent) and nitrogen. Temperatures average about 900°F at the surface, and atmospheric pressures run some 100 times those on Earth. Theories about the planet abound: that it has 15-mile-high mountains, for example, or that its clouds could provide a habitat for some forms of life—perhaps hydrogen-filled floating bladders that consume water vapor and minerals upon direct contact, like our air plants. Such organisms, it is said, could have evolved from surface life forms early in the history of Venus and could operate entirely on known terrestrial principles. Only time and further exploration of Venus can provide the answers to this type of speculation.

The United States is planning probes of Mercury, the innermost planet, using Mariner-type spacecraft to visit both Venus and Mercury on the same mission. Top priority is being given to investigation of the atmosphere and ultraviolet spectrum of Venus, the surface characteristics of Mercury, and the interplanetary medium en route. The United States also is collaborating with West Germany on a separate Helios project to explore space some 10 million miles inside the orbit of Mercury.

Scientific interest in the asteroids—the minor planets orbiting primarily between Mars and Jupiter—derives from their potential for providing insights into the origin of the solar system through analyses of their age, chemical makeup, and structure. It has been estimated that there are at least 80,000 asteroids, but only about 1,650 have been numbered and had their orbits calculated. Ceres, the largest, has a diameter of about 480 miles, but most are much smaller. Their total mass is probably less than that of the Moon. It has been suggested that asteroids might make "hitchhiking" bases for interplanetary travel, and that their material might be used for rocket shielding, supplies, and propellants. In any event, more knowledge about the asteroids is needed if space travel to the outer planets is to become commonplace.

Of the planets in the outer solar system, Jupiter is the most fascinating. It has 12 satellites, a mass some 318 times that of the Earth, a rapid rotation period (less than 10 Earth hours), and an abundance of hydrogen and helium in its atmosphere. Jupiter is more like a star than a planet, in the sense that it appears to emit twice as much energy as it absorbs. For this reason, scientists have called it "the Rosetta stone of the solar system."

Spacecraft missions are planned to explore the interplanetary medium and cosmic radiation beyond the orbit of Mars, to investigate possible spacecraft hazards from asteroids, and to measure various properties of Jupiter, including its cloud structure, thermal balance, magnetic fields, radiation belts, and electromagnetic emissions. A number of photographs are also planned during these two-year missions covering almost 500 million miles.

To reach the four outermost planets—Saturn, Uranus, Neptune, and Pluto—a series of "Grand Tours" are planned for the late 1970s and 1980s, when these planets will be aligned in such a manner that probes can use Jupiter as a pivot, picking up momentum from its gravitational attraction. Consequently these missions will be able to reach all the outer planets in 9 years, rather than the 40 years that otherwise would be required to journey to Pluto. This extraordinary alignment was last observed in the days when Thomas Jefferson was President, and will not reoccur for another 179 years.

Many advances will have to be made in the design and reliability of spacecraft systems, especially in the computer area, in order to carry out missions of so many years' duration. The on-board computer must be able to diagnose malfunctions, then repair them, without waiting for instructions from the Earth. This capability is necessary because of the distances involved: in the time that it would take a warning signal to reach the Earth and an instruction to be sent back to the probe, the original malfunction might be followed by a sequence of irredeemable events, causing the mission to fail. One plan is to divide the computer into three lobes; if one lobe malfunctioned, it would be "outvoted" by the other two, its memory erased and then restored to conform to the others. If the lobe continued to malfunction, its connections would be severed and the journey completed under the command of the two agreeing lobes. The potential applications of such advanced computer concepts for earthbound tasks herald yet other dividends from the space program.

COSMIC PHYSICS AND ASTRONOMY

Most of the many satellites and probes launched since Sputnik initiated the space age in 1957, have been designed to broaden our understanding of interplanetary space (by direct measurements) and of intergalactic space (by improved observations). At the same time, by carefully observing and monitoring satellites in orbit, scientists also have pieced together a completely new picture of the Earth's upper atmosphere, with its strange pulsations, motions, and exotic chemical reactions. Airglow, polar auroras, and the relationships between geomagnetic disturbances and the plasma erupting from the Sun's surface have been carefully studied.

Cosmic rays—highly penetrating radiation from extraterrestrial sources—continuously bombard the Earth. A small percentage of the primary particles come from the Sun (solar cosmic rays). The higher-energy particles come from outside the solar system (galactic cosmic rays), with most originating within our own Milky Way galaxy. It is theorized that the major source of these cosmic rays is supernova explosions, which periodically occur and accelerate electrons to very high energy levels. Large magnetic fields which further accelerate the cosmic rays are assumed to exist between the stars and around gas clouds.

Most of the cosmic rays that slice through the atmosphere impact with atoms, thus creating secondary cosmic rays. To determine the sources and flow directions of primary cosmic radiation, it is necessary to establish a monitor station high above the atmosphere—and possibly beyond the solar system itself. Scientists have theorized that, just as the Earth's atmosphere protects us from solar ultraviolet radiation, so the Sun's plasma, or solar wind, protects us from intense cosmic radiation, particularly during the peak periods of activity in the 11-year solar cycle. Cosmic knowledge of this type can well be applied for a better understanding of ourselves and out host planet and to predict and compensate for the forces which affect us.

It also has been proposed that physicists engaged in elementary-particle research take advantage of the extraordinary high energies of cosmic rays by using satelliteborne detectors in place of ground-based cyclotrons and accelerators. A satellite system could provide about 200 times the energy of the largest existing accelerator and at considerably less cost. Experiments for such a satellite system would include measurement of the cosmic-ray energy spectrum, searches for antiparticles and superheavy nuclei, as well as the production of heavy secondary particles from impacts with the extremely high-energy cosmic-ray protons.

The Sun, of course, is only one star among 100 billion in a single galaxy, within a universe containing an untold number of galaxies. It was with this tremendous vista in mind that the NASA Astronomy Missions Board, in making its recommendations for the 1970s, emphasized the need for scientific progress "on such problems as the origin of the universe; the course of stellar evolution, including the ultimate destiny of the Sun and Solar System; the existence of other planetary systems, some of which may support other forms of intelligent life; and other problems with great philosophical significance."

One of the key elements of the astronomical-satellite program is the Orbiting Astronomical Observatory (OAO), which has been ranked with the invention of the telescope in its importance to astronomy. This project has accomplished tremendous mapping efforts, locating many galaxies and stars with extremely bright ultraviolet radi-

The Orbiting Astronomical Observatory has been ranked with the invention of the telescope in its importance to astronomy. In its first month of operation, OAO-1 obtained 20 times as many ultraviolet data on stars as had been gained during the previous 15 years by rocketborne instruments. (NASA)

ations. Before the OAO, the expenditure of some 40 sounding rockets over a 15-year period was required to obtain just 3 hours of ultraviolet data from 150 stars. The first successful OAO satellite, in its initial month of operation, gained 20 times as much information.

The results of the OAO program are helping astronomers to understand the processes which create stars and galaxies. Many young stars with surface temperatures of 45,000°F (as compared with the 10,000°F of our middle-aged Sun) have been found. Exciting new data concerning bright quasars and colliding star systems, as well as greater knowledge of the Earth's atmosphere—based on observations of stars as they "set"—have been realized from the OAO instruments. Distances between galaxies might even be determined with sufficient accuracy to yield the scale and curvature of the universe. Supernova explosions and the mysterious pulsars—gigantic sources of energy which emit radiation at short, precise intervals—are being surveyed, as well as strange, newly discovered naturally occurring laser processes having tremendous energy potentials.

Programs such as the OAO and others planned by NASA in space astronomy promote the unification of science as they involve the application of talent from many different disciplines. The data gathered from these programs are of keen interest not only to astronomers but to chemists, mathematicians, theoretical physicists, geologists, and geophysicists. Space instrumentation developed for research in the solar, planetary, X-ray and gamma-ray, ultraviolet, infrared, and radio fields also has potential applications elsewhere, as precision measurement techniques and data reduction methods may be readily adapted to other basic and applied investigations. And man on Earth will gain accordingly from his explorations of the universe.

MATERIALS AND CHEMICAL PROCESSING

Manufacturing processes under less-than-1G conditions have been practiced since the Middle Ages, when shot was produced by pouring liquid lead through a screen to form droplets during free fall. Reflecting glass spheres for traffic signs and hollow glass beads are made by a similar process today. Free fall at zero gravity has been extended to about 20 seconds with special airplane flight maneuvers. But these relatively few seconds of reduced gravity on the Earth cannot match the long-term zero gravity of an orbiting station and the potential it has for manufacturing processes.

The potential of zero gravity for extensive processing of materials has been compared with the development of vacuum processes during the seventeenth century. The steam engine, the internal combustion engine, gas liquefication, radio—all critical to our current technology—are heavily dependent on the utilization of vacuum and pressure-differential systems.

For the near future (through 1985), Earth-orbital space stations and lunar bases are planned from which we can learn more about the space environment and its potential uses. Scientific and technological experiments suitable for stations include those investigating fluid systems, materials processing, and the role of gravity on the functions of plants and animals. From an economic viewpoint, the materials-processing area appears to have the greatest potential. And materials development has historically been the key to advancing the general state of the technological arts.

The materials-processing and manufacturing operations now used on Earth have been reviewed by NASA's Marshall Space Flight Center for potential applications to manufacturing in space. Hans F. Wuenscher, of the center's Manufacturing Engineering Laboratory, divides space manufacturing processes into three categories, as follows:

1. Application of existing terrestrial processes for space operations and manufacturing.
   EXAMPLES: Diffusion bonding
   Electron beam welding
   Exothermic tube brazing
   Laser welding, cutting, and drilling
   Vapor deposition
2. Optimization of terrestrial manufacturing processes to the space environment.
   EXAMPLES: Crystal and whisker growing
   Levitation melting
   Precision heat treatment
   Thin film processing
   Vacuum melting and casting
3. Development of unique space manufacturing processes.
   EXAMPLES: Blending of materials of different density
   Powder conversion
   Composite casting
   Cohesion casting

The third category is being investigated in considerable depth by NASA and industrial contractors, with emphasis on both technological and economic considerations. A more detailed analysis of this category has resulted in a listing of unique space processes and their potential product applications. Of course, it is important that the value of these potential products exceed the costs of space transportation and processing, if these research efforts are to be economically viable.

TABLE I

SPACE PROCESSES AND APPLICATIONS

SPACE PROCESS EMPLOYING REDUCED GRAVITY	POTENTIAL APPLICATIONS
Buoyancy and thermal convection-free processes:	
Blending of various materials in nonmetallic matrix	Radiation and thermal shielding for electronic packages
Composite casting by dispersion of compounds and fibers in metallic matrix	High-strength, heat-resistant structures and alloys
Powder conversion into castings	Isotopes; nuclear fuels
Cohesion- and adhesion-controlled processes (using molecular forces):	
Adhesion casting by floating layers of materials inside or outside a mold	Multilayer coated isotopes; optical and solid state components
Blow casting using gas inclusion and inspection and die cutouts	Hollow spheres; ultrathin membranes; complex hollow bodies and multilayer structures
Cohesion, or surface-tension, casting (also by using inertia or electrostatic fields)	Ball bearings; hybrid computer components; large optical blanks for improved lenses
Controlled-density casting by control of bubble distribution	Foamed and buoyant materials; variable-density turbine blades; armor plate

NASA is now considering research programs in various areas, ranging from the manufacture of ball bearings and foamed materials to the analysis of crystallization and gaseous reactions in outer space. Ball bearings, for example, are normally produced by feeding a length of wire between two hemispherical dies, filling the dies which then form a flange around the ball. A series of centerless grinders, each of a finer grade, then reduce surface roughness and diameter variation to the microinch level or further. Hollow ball bearings, having less

Combining tiny fibers with molten metal in a zero-gravity environment results in a low-weight, high-strength composite material, one of the many entirely new products that could be produced in orbiting "space factories." (GENERAL DYNAMICS)

mass, offer particular advantages at high speeds because of the reduction in centrifugal force and sliding friction. They are conventionally manufactured by welding two hemispherical shells together; but the welding process itself creates instability and possible fracture points, due to density variations and interior voids at the interface of the base material and weld joint.

In the space environment, experiments are planned to investigate the possibilities of manufacturing both solid and hollow ball bearings cast in structures unique in metallurgy, with finer grain, less surface roughness, greater hardness, and more perfectly spherical shape. These improvements over terrestrial ball-bearing production could help to revolutionize our machine-based civilization by reducing friction and noise caused by ball bearings. In zero gravity, even the elusive perpetual motion machine may become practical as friction disappears and bearing loads are reduced to zero.

Experiments with composite materials will also be conducted to investigate the potential for more uniform structures. To determine if the characteristics of composite materials can be improved, reinforcing matter such as fibers and dispersed particles will be introduced into the parent substance. (A simple experiment of this nature was conducted during the return flight of Apollo 14 from the Moon.) Adhesion and cohesion become much more important in the zero-gravity mixing process, and surface-tension and capillary forces can be put to work in this unique environment to develop improved structural characteristics.

Foamed materials with closed (sealed) microscopic cells are being considered as candidates for unique space manufacturing products. Conventional lightweight metallic foams now being made usually have porous cells and low structural load-carrying properties. In mixing gas in varying amounts with liquid metals in the space process, the gas bubbles should not rise to the surface as they do on the Earth—and the resulting foamed material could provide unique characteristics for extremely strong but lightweight structures. With the development of micropore structures in space, novel applications such as lightweight armor plate, variable-density rotating components, and insulation materials with varying temperature gradients can be conceived.

We do not fully understand how crystals (such as in gems or metals) arrange themselves in small geometric patterns, called lattices, and repeatedly form the same patterns as they grow. It does appear that the formation and growth of crystals may be different in the zero-gravity environment. Slow motion pictures are planned to study dislocations or defects in crystals, caused by crystal-seed inheritance as well as thermal and other stresses. These defects, which limit crystal strength, may be significantly reduced or even eliminated in the zero-gravity environment to create a much more nearly perfect crystal. Material strength may be increased by a whole order of magnitude. In addition, dislocation-free fibers for reinforcing composite materials, and improved materials for thermal control may result from these studies, thus expanding the world's inventory of industrial materials, raising the quality of goods, and increasing the potential for a higher standard of living on Earth through development of more efficient products.

The discipline of chemistry has been proposed as a special candidate for experimental research in space. For example, the effect of gravity upon the kinetic nature of chemical reactions is not readily apparent. One view is that the forces involved in chemical bonding are so large in comparison with the force of gravity that any effect of gravity is quite negligible; hence no change should be expected under conditions of zero gravity. However, gravity could be a hidden variable in chemical reactions which has so far been ignored because of its all-pervading presence on Earth. Benefits may accrue from such research in several areas–including the exploration and discovery of new and unusual states of matter in space, and accumulation of new knowledge concerning the distribution of the chemical elements –which could aid substantially in the study of cosmology.

Gaseous reactions might be easier to analyze under zero gravity. No experimental difficulty is anticipated, as a gas fills its container completely and uniformly through kinetic energy diffusion. It is likely that such experiments could be carried out in space with conventional glassware and other equipment. Studies of liquid reactions, however, may be difficult, as liquid media are not readily poured in space and have a tendency to break apart into suspended droplets.

Advanced space stations could accommodate modules providing a shirt-sleeve environment for the astronauts, with maintenance facilities, compartments for raw materials, and chambers for manufacturing and processing activities. A module, perhaps weighing as much as 16 tons and containing test apparatus, monitoring equipment, closed-circuit television systems, would prove an extremely flexible research facility for Earth-orbital manufacturing operations. By the mid-1970s, commercially attractive processes in Earth orbit should have been identified so that industry can begin to consider their realization.

LEARNING ABOUT LIFE

As man extends his efforts to explore space, he also continues his quest for complete biological comprehension. Living organisms, from microbe to man, are extremely sensitive to their environment, and space provides unique environmental conditions to which life forms can be exposed. Already many biomedical, behavioral, and life support studies have been conducted to aid man's venture into space. Additional studies are now being considered to explore how the weightless environment available in

Earth orbit can be used to increase man's understanding of life processes, of his own interactions with his environment, and of his tolerance levels.

A by-product of the biological studies necessitated by the space program has been the incorporation of space research data and technology into terrestrial clinical medicine, which in turn has inspired new multidisciplinary approaches to diagnosis and treatment. The merger of the efforts of medical doctors and engineers has led not only to better communication and cooperation between the two groups but to improved understanding of dynamic systems, including the biological, physical, chemical, and electrical, and to the development of such new analytical tools as telemetry and the microcircuit. As a result, important discoveries are being made which will assist in improving substantially the medical care of earthlings.

The basic goals of life science research within the space program are as follows:

1. Obtain an understanding of man's reaction to long-term exposure in the space environment.
2. Advance the technology for support of man during extended exposure to the space environment.
3. Study effects of the space environment on living organisms to increase the understanding of life on Earth.

Attaining the first objective is essential to future manned activities in space and is the initial objective of the Skylab space station program. Development of effective life support systems—the second goal—is another key factor in all manned space research programs and deserves its high priority. Finally, a better understanding of the basic nature and dynamics of normal life processes should be obtained by removing certain terrestrial environmental stimuli, especially the normal gravitational force; modifying exposure to cyclic geophysical conditions; and then observing organic response to the new conditions.

The completely sterile environment afforded by the space medium presents exciting possibilities for the development of pure vaccines, and zero gravity possibly offers a rapid growth potential for these cultures. The pharmaceutical industry could take advantage of frictionless centrifuges in space to separate the molecules of complicated drugs having close to the same density. These possibilities may eventually result in exotic formulas for "miracle" drugs at a cost that even the developing nations can afford.

Beyond the experimental phase of life science research in Earth orbit, a number of interesting applications and long-term objectives have been proposed for space stations. These include orbital hospitals offering surgical and therapeutic treatments unique to the zero-gravity environment. Partial paralytic conditions, trophic ulcers due to pressure, cardiovascular problems due to inadequate heart-pumping pressure, and burns where skin-to-bed contact is not desirable may be especially susceptible to treatment in the space environment. Physical therapy could take on new dimensions, with patients learning to use their limbs again in an environment where the gravitational force is gradually increased. Certainly the design and operation of hospitals in space, with their completely closed environments and automated systems, could provide significant knowledge for use in their terrestrial counterparts, especially in the area of ecological control.

POWER WITHOUT POLLUTION

One of the potentially most useful applications of space technology is the realization of "clean" power from the Sun. As envisioned by Dr. Peter Glaser of Arthur D. Little, Inc., this source of energy is urgently needed on a large scale as the Earth's finite reserves of fossil fuels decline, as air pollution increases, and as thermal pollution from nuclear power plants threatens water sources. Disposal of radioactive waste materials from such plants also poses a formidable problem. Increased demands for power necessitate a search for new energy sources, such as the solar conversion system, that will be free of adverse effects on our environment.

Solar cells, one of the most useful items refined by the space program, have been largely responsible for the success of many satellites and space probes. Unfortunately they are not now economically effective on the Earth, due to atmospheric obstructions and because they are limited to daytime exposure. However, other solar energy devices, such as solar stills, water heaters, engines, and furnaces, have in the past been used on Earth with various degrees of success.

To meet the energy needs of the future, Dr. Glaser conceives of huge Earth-synchronous satellites at an altitude of 22,300 miles above the Earth, with highly efficient solar-energy conversion systems aboard large-area collectors facing the Sun. The electricity generated by these collectors would be directed to amplifiers, which would then convert it to microwave radiations to be beamed to the Earth. In order to insure safe operation, the radiating antenna would be carefully guided to transmit the power to receiving stations at a level of about 0.01 watt per square centimeter—or one-tenth the power density of solar radiation normally received on the Earth. A 10-cm wavelength has been selected to minimize atmospheric absorption. The receiving stations would efficiently reconvert the beamed microwave energy to electricity, then feed it through transmission lines to surrounding communities.

A similar system could be used to provide power for the space stations and orbital modules now being planned. Man's efficient use of solar energy can thus gain him unlimited energy for the future without using up the Earth's resources. Such cheap, clean power would ease many of the most pressing problems associated with exploding population and industrial growth.

EXTRATERRESTRIAL RESOURCES

Man's desire to explore the planets is considerably influenced and stimulated by his recent discoveries and new knowledge of the Moon's natural resources. Other extraterrestrial resources, such as minerals or chemical compounds developed by forces now only vaguely understood, may eventually lure colonies of our species forth to search for, recover, and use the materials of these other worlds.

The solar system contains 35 known bodies of more than 100 miles in diameter and thousands of smaller bodies, mainly asteroids, with unknown resources. Domed or subsurface bases could be employed to facilitate their exploration and, if justified by discovery of valuable raw materials, their colonization. One of the obvious potential contributions of space colonies will be the required development of

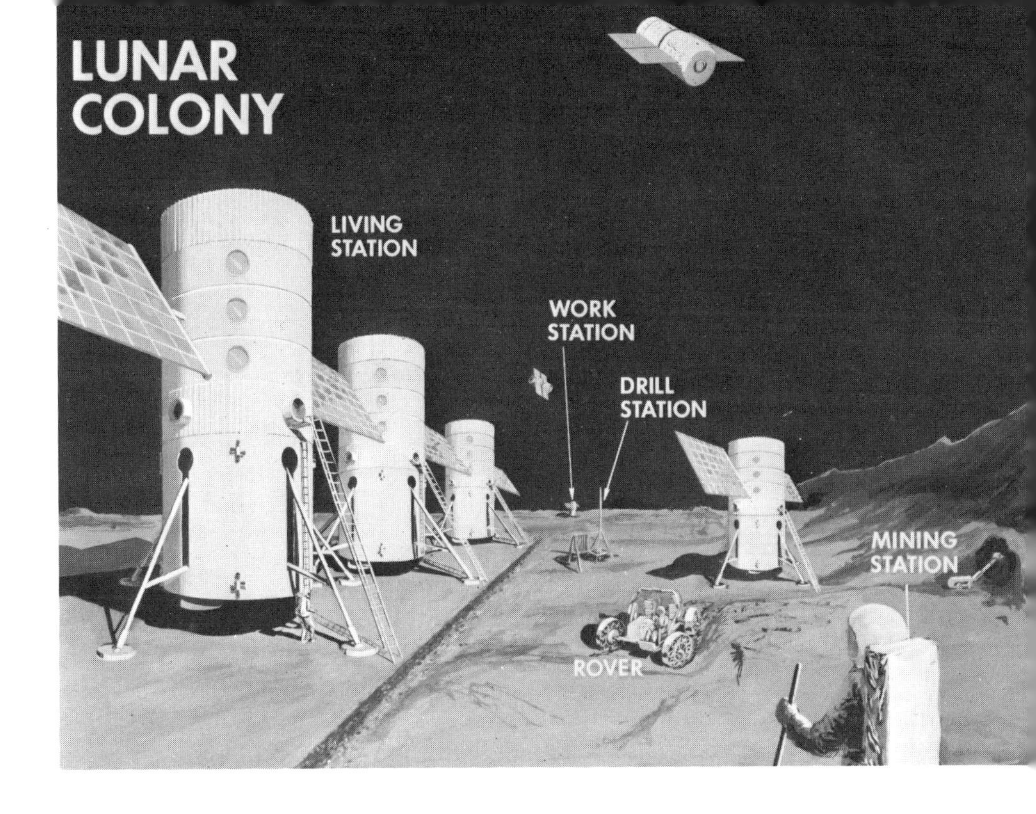

Artist's conception of a lunar colony, making use of space station modules as shelters for men. In the more distant future, large domed bases, as below, might be established on the Moon for mining, manufacturing, and scientific operations. (NASA)

efficient closed environmental systems and the ecological analysis so necessary for them. These efforts are desperately needed to help solve the Earth's population and pollution problems. Likewise, the harmonious coexistence required of individuals within their enclosed system in the alien surroundings of space, could serve to establish guidelines for earthbound society. Closed-cycle systems, recycling and exchange systems, and efficient utilization of all energy processes also would be vastly improved in the course of space research and colonization.

Before extraterrestrial bases can be established, it will be necessary to search for, and learn to use, the extraterrestrial substances needed for constructing and operating the bases. This will hold true also for most materials processed in space, since it is necessary to keep to a minimum the bulk of material which must be transported from the Earth. Surveys, exploration, and sampling, conducted through experimental programs using both unmanned and manned vehicles, are being considered for the initial search effort. Depending on the technological and economic feasibility, further efforts will then be made to establish bases for more extended study. One representative scientific benefit that might be expected from such bases is the installation of huge radio telescopes on the far side of the Moon, with consequent protection from Earth-generated noise.

After specific data have been collected concerning extraterrestrial materials of value, unique mining systems may be developed. Scraping, drilling, blasting, and other conventional mining methods probably will be considered. However, the use of entirely new methods of extraction can be anticipated, owning to the alien environment and remoteness of the sites. New sources of power and power distribution systems can also be expected—with potential application for our needs here on Earth.

Altering the environment of other planets to make them more earthlike (terraforming), by freeing their oxygen and water with huge energy sources, has been considered as a mode of development for human habitation. In any event, the colonization of other worlds can be considered as an insurance policy allowing for the survival of our species, regardless of our reckless behavior at home base.

Epilogue

A LETTER TO ZAMBIA

IN THE SPRING OF 1970, SISTER MARY Jucunda, O.P., wrote a letter from Kabwe, Zambia, to the famed physicist and space scientist Ernst Stuhlinger. In it, she asked how one could justify spending billions of dollars on astronautical endeavors while millions of children are starving here on Earth. It was a compassionate question, one that is by no means uncommon and one that deserves an answer. What is curious, however, is that such social problems as hunger, crime, urban decay, and pollution are so often introduced with the implicit or explicit suggestion that if the space program were to be dismantled, the social evil would somehow be alleviated.

Other major activities, including the cosmetic and tobacco industries, involve greater expenditures by Americans than the $3.5 billion allocated annually to NASA for its diverse programs. These industries may be criticized as catering to human vanity and damaging to human health, but it is rarely suggested that the efforts and monies they represent be channeled into, for example, urban renewal. That space activities should be more vulnerable in the scale of values than lipstick and the kingsize filter tip is more melancholy than surprising.

A sincere, practical, and eloquent answer to Sister Jucunda's question was framed by Dr. Stuhlinger in early May 1970. He wrote as follows:

Dear Sister Mary:

Your letter was one of many which are reaching me every day, but it has touched me more deeply than all the others because it came so much from the depth of a searching mind and a compassionate heart. I will try to answer your question as best I possibly can.

First, however, I would like to say what great admiration I have for you, and for all your many brave sisters, because you are dedicating your lives to the noblest cause of man: help for his fellow men who are in need.

You asked in your letter how I can suggest the expenditures of billions of dollars for a voyage to Mars, at a time when many children on this Earth are starving to death. I know that you do not expect an answer such as, "Oh, I did not know that there are children dying from hunger, but from now on I will desist from any kind of space research until mankind has solved that problem!" In fact, I have known of famined children long before I knew that a voyage to the planet Mars was technically feasible. However, I believe, like many of my friends, that travelling to the Moon and eventually to Mars and to other planets is a venture which we should undertake now, and I even believe that this project, in the long run, will contribute more to the solution of these grave problems we are facing here on Earth than many other potential projects of help which are debated and discussed year after year, and which are so extremely slow in yielding tangible results.

Before trying to describe in more detail how our space program is contributing to the solution of our earthly problems, I would like to relate briefly a true story which may help support the argument. About four hundred years ago, there lived a count in a small town in Germany. He was one of the benign counts, and he gave a large part of his income to the poor in his town. This was much appreciated, because poverty was abundant during medieval times, and there were epidemics of the plague which ravaged the country frequently. One day, the count met a strange man. He had a workbench and little laboratory in his house, and he labored hard during the daytime so that he could afford a few hours every evening to work in his laboratory. He ground small lenses from pieces of glass; he mounted the lenses in tubes, and he used these gadgets to look at very small objects. The count was particularly fascinated by the tiny creatures that could be observed with the strong magnification, and which nobody had ever seen before. He invited the man to move with his laboratory to the castle, to become a member of the count's household, and to devote henceforth all his time to the development and perfection of his optical gadgets as a special employee of the count. The townspeople, however, became angry when they realized that the count was wasting his money, as they thought, on a stunt without purpose. "We are suffering from the plague," they said, "while he is paying that man for a useless hobby!" But the count remained firm. "I give you as much as I can afford," he said, "but I will also support this man and his work, because I know that someday something good will come out of it!" Indeed, something very good came out of his work, and also out of similar work done by others at other places: the microscope. It is well known that the microscope has contributed more than any other invention to the progress of medicine, and that the elimination of the plague and many other contagious diseases from most parts of the world is largely a result of studies which the microscope made possible. The count, by retaining some of his spending money for research and discovery, contributed far more to the relief of human suffering than he could have contributed by giving all he could possibly spare to his plague-ridden community.

The situation which we are facing today is similar in many respects. The President of the United States is spending about 200 billion dollars in his yearly budget. This money goes to health, education, welfare, urban renewal, highways, transportation, foreign aid, defense, conservation, science, agriculture, and many installations inside and outside the country. About 1.6 per cent of this national budget was allocated to space exploration this year. The space program includes Project Apollo, and many other smaller projects in space physics, space astronomy, space biology, planetary projects, Earth resources projects, and space engineering. To make this expenditure for the space program possible, the average American taxpayer with 10,000 dollars income per year is paying about 30 dollars for space. The rest of his income, 9,970 dollars, remains for his subsistence, his recreation, his savings, his taxes, and all his other expenditures.

You will probably ask now: "Why don't you take 5 or 3 or 1 dollar out of the 30 space dollars which the average American taxpayer is paying, and send these dollars to the hungry children?" To answer this question, I have to explain briefly how the economy of this country works. The situation is very similar in other countries. The government consists of a number of departments (Interior; Justice; Health, Education and Welfare; Transportation; Defense; and others), and of Bureaus (National Science Foundation; National Aeronautics and Space Administration; and others). All of them prepare their yearly budgets according to their assigned missions, and each of them must defend its budget against extremely severe screening by congressional committees, and against heavy pressure for economy from the Bureau of the Budget and the President. When the funds are finally appropriated by Congress, they can be spent only for the line items specified and approved in the budget. The budget of the National Aeronautics and Space Administration, naturally, can contain only items directly related to aeronautics and space. If this budget were not approved by Congress, the funds proposed for it would not be available for something else; they would simply not be levied from the taxpayer, unless one of the other budgets had obtained approval for a specific increase which would then absorb the funds not spent for space. You may realize from this brief discourse that support for hungry children, or rather a support in addition to what the United States is already contributing to this very worthy cause in the form of foreign aid, can be obtained only if the appropriate department submits a budget line item for this purpose, and if this line item is then approved by Congress.

You may ask now whether I personally would be in favor of such a move by our government. My answer is an emphatic yes. Indeed, I would not mind at all if my annual taxes were increased by a number of dollars for the purpose of feeding hungry children wherever they may live.

I know that all of my friends feel the same way. However, we could not bring such a program to life merely by desisting from making plans for voyages to Mars. On the contrary, I even believe that by working for the space program I can make some contribution to the relief and eventual solution of such grave problems as poverty and hunger on Earth. Basic to the hunger problem are two functions: the production of food, and the distribution of food. Food production by agriculture, cattle ranching, ocean fishing, and other large-scale operations is efficient in some parts of the world, but drastically deficient in many others. For example, large areas of land could be utilized far better if efficient methods of watershed control, fertilizer use, weather forecasting, fertility assessment, plantation programming, field selection, planting habits, timing of cultivation, crop survey, and harvest planning were applied. The best tool for the improvement of all these functions, undoubtedly, is the artificial Earth satellite. Circling the globe at a high altitude, it can screen wide areas of land within a short time, it can observe and measure a large variety of factors indicating the status and conditions of crops, soil, droughts, rainfall, snow cover, etc., and it can radio this information to ground stations for appropriate use. It has been estimated that even a modest system of Earth satellites equipped with Earth resources sensors, working within a program for world-wide agricultural improvements, will increase the yearly crops by an equivalent of many billions of dollars.

The distribution of the food to the needy is a completely different problem. The question is not so much one of shipping volume; it is one of international cooperation. The ruler of a small nation may feel very uneasy about the prospect of having large quantities of food shipped into his country by a large nation, simply because he fears that along with the food there may also be an import of influence and foreign power. Efficient relief from hunger, I am afraid, will not come before the boundaries between nations have become less dividing than they are today. I do not believe that space flight will accomplish this miracle overnight. However, the space program is certainly among the most promising and powerful agents working in this direction. Let me only remind you of the recent near-tragedy of Apollo 13. When the time of the crucial reentry of the astronauts approached, the Soviet Union discontinued all Russian radio transmissions in the frequency bands used by the Apollo Project in order to avoid any possible interference, and Russian ships stationed themselves in the Pacific and the Atlantic Oceans in case an emergency rescue should become necessary. Had the astronaut capsule touched down near a Russian ship, the Russians would undoubtedly have expended as much care and effort in their rescue as if Russian cosmonauts had returned from a space trip. If Russian space travellers should ever be in a similar emergency situation, Americans would do the same without any doubt.

Higher food production through survey and assessment from orbit, and better food distribution through improved international relations, are only two examples of how profoundly the space program will impact life on Earth. I would like to quote two other examples: stimulation of technological development, and generation of scientific knowledge.

The requirements for high precision and for extreme reliability which must be imposed upon the components of a Moon-travelling spacecraft are entirely unprecedented in the history of engineering. The development of systems which meet these severe requirements has provided us a unique opportunity to find new materials and methods, to invent better technical systems, to improve manufacturing procedures, to lengthen the lifetimes of instruments, and even to discover new laws of nature. All this newly acquired technical knowledge is also available for application to Earth-bound technologies. Every year, about a thousand technical innovations generated in the space program find their ways into our Earthly technology where they lead to better kitchen appliances and farm equipment, better sewing machines and radios, better ships and airplanes, better weather forecasting and storm warning, better communications, better medical instruments, better utensils and tools for everyday life. Presumably, you will ask now why we must develop first a life support system for our Moon-travelling astronauts, before we can build a remote-reading sensor system for heart patients. The answer is simple: significant progress in the solution of technical problems is frequently made not by a direct approach, but by first setting a goal of high challenge which offers a strong motivation for innovative work, which fires the imagination and spurs men to expend their best efforts, and which acts as a catalyst by inducing chains of other reactions. Space flight, without any doubt, is playing exactly this role. The voyage to Mars will certainly not be a direct source of food for the hungry. However, it will lead to so many new technologies and capabilities that the spin-offs from this project alone will be worth many times the cost of its implementation.

Besides the need for new technologies, there is a continuing great need for new basic knowledge in the sciences if we wish to improve the conditions of human life on Earth. We need more knowledge in physics and chemistry, in biology and physiology, and very particularly in medicine to cope with all these problems which threaten man's life: hunger, disease, contamination of food and water, pollution of the environment. We need more young men and women who choose science as a career, and we need better support for those scientists who have the talent and the determination to engage in fruitful research work. Challenging research objectives must be available, and sufficient support for research projects must be provided. Again, the space program with its wonderful opportunities to engage in truly magnificent research studies of the Moon and planets, of physics and astronomy, of biology and medicine is an almost ideal catalyst which induces the reaction between the motivation for scientific work, opportunities to observe exciting phenomena of nature, and material support needed to carry out the research effort.

Among all the activities which are directed, controlled, and funded by the American government, the space program is certainly the most visible, and probably the most debated activity, although it consumes only 1.6 per cent of the total national budget, and 3 per mill of the gross national product. As a stimulant and catalyst for the development of new technologies, and for research in the basic sciences, it is unparalleled by any other activity. In this respect, we may even say that the space program is taking over a function which for three or four thousand years has been the sad prerogative of wars. How much human suffering can be avoided if nations, instead of competing with their bomb-dropping fleets of airplanes and rockets, compete with their Moon-travelling spaceships! This competition is full of promise for brilliant victories, but it leaves no room for the bitter fate of the vanquished which breeds nothing but revenge and new wars.

Although our space program seems to lead us away from our Earth and out toward the Moon, the Sun, the planets, and the stars, I believe that none of these celestial objects will find as much attention and study by space scientists as our Earth. It will become a better Earth, not only because of all the new technological and scientific knowledge which we will apply to the betterment of life, but also because we are developing a far deeper appreciation of our Earth, of life, and of man. The photograph which I enclose with this letter shows a view of our Earth as seen from Apollo 8 when it orbited the Moon at Christmas, 1968. Of all the many wonderful results of the space program so far, this picture may be the most important one. It opened our eyes to the fact that our Earth is a beautiful and most precious island in an unlimited void, and that there is no other place for us to live but the thin surface layer of our planet, bordered by the bleak nothingness of space. Never before did so many people recognize how limited our Earth really is, and how perilous it would be to tamper with its ecological balance. Ever since this picture was first published, voices have become louder and louder, warning of the grave problems that confront man in our times: pollution, hunger, poverty, urban living, food production, water control, overpopulation. It is certainly not by incident that we begin to see the tremendous tasks waiting for us at a time when the young space age has provided us the first good look at our own planet. Very fortunately, though, the space age not only holds out a mirror in which we can see ourselves, it also provides us with the technologies, the knowledge, the challenge, the motivation, and even the optimism to attack these tasks with confidence. What we learn in our space program, I believe, is fully supporting what Albert Schweitzer had in mind when he said: "I am looking at the future with concern, but with good hope."

My very best wishes will always be with you.

Very sincerely yours,
Ernst Stuhlinger
Associate Director for Science

Appendix

INFORMATION SOURCES FOR SPACE TECHNOLOGY

THE SPACE PROGRAM HAS INITIATED AND helped to discover many new things during its brief existence. These range from the expansion of scientific knowledge about our solar system, the universe, and ourselves, to advancing the state-of-the-art for materials and machines. The accumulated results of aerospace investigations have been carefully collected, summarized, indexed, and stored for dissemination to potential users by the National Aeronautics and Space Administration, as specified in its charter.

An information bank of more than 400,000 documents has been accumulated by the scientific and technical information system of NASA. These documents include government, industry, and academic reports, articles, and reviews from all over the world. About one-third came from foreign countries, including Russia and those of Eastern Europe. Approximately 6,000 additions are made to the collection each month.

Each document is examined upon receipt by NASA, given an accession number, and, normally, converted to 4- by 6-inch microfiches for ease of storage and handling. Trained indexers and abstracters then process the material for automatic use by high-speed computers. Thus the information can be quickly located by its accession number, corporate source, authorship, or subject matter. Abstract journals are also prepared under NASA sponsorship, for users of the information bank to keep abreast of current developments and as a retrospective research tool.

Anyone can subscribe to the abstract journals: *Scientific and Technical Abstract Reports* (STAR) and *International Aerospace Abstracts* (IAA). Both are issued twice monthly. Also, *Selected Current Aerospace Notices* (SCAN), based on NASA information bank computer searches, provide bibliographic and abstract data on approximately 200 specific scientific and technical topics. These notifications present new findings in specialized areas such as aerospace medicine, laminated materials, ultrasonic testing, etc., and assure the user that his current interests are kept up to date. Computer searches of the NASA information bank can also be made to meet a user's particular needs.

Besides the organizational and dissemination services related to scientific and technical information, NASA also generates its own formal publications, including technical memorandums, translations, and contractor reports. Special publications and reports are also produced as program summaries, conference proceedings, monographs, handbooks, and bibliographies. A general series is published for matter of broad interest such as space photography and satellite information, and other series include histories, chronologies, and management information.

Perhaps the most important documents related to the subject of this book are issued by the NASA Office of Technology Utilization. The most broadly used document is a one- or two-page technical announcement called a tech brief, which lists innovations, techniques, processes, concepts, and devices sponsored by NASA. Some of these notices are also issued jointly with the Atomic Energy Commission, concerning work that one or both agencies have sponsored, and are called AEC-NASA tech briefs. Organized by specific categories, these announcements also list additional reference sources for the benefit of the user.

Special reports and surveys are generated under the NASA Technology Utilization (TU) program to facilitate the transfer of aerospace technology to other areas of application. Institutions known as Regional Dissemination Centers, sponsored by NASA, help to disseminate aerospace information among industry and organizations not directly or normally in contact with this type of technology. These nonprofit research institutions and university-based facilities were established as field offices at the local level to identify problems, to supply pertinent documents, to perform researches, and, hopefully, to develop evaluation programs and provide transfer assessment and feedback to NASA. Each regional center is responsive to a specific geographic and economic environment; hence their services and their fees vary. The addresses and telephone numbers of the centers are given below, and a prospective client may consult any of them about its offerings and charges.

Aerospace Research Applications Center
Indiana University Foundation
Bloomington, Indiana 47405
Phone (812) 337-7970

Knowledge Availability Systems Center
University of Pittsburgh
Pittsburgh, Pennsylvania 15213
Phone (412) 621-3500, ext. 6352

Technology Application Center
University of New Mexico
Box 185
Albuquerque, New Mexico 87106
Phone (505) 277-3118

New England Research
Application Center
University of Connecticut
Storrs, Connecticut 06268
Phone (203) 429-6616

North Carolina Science and
Technology Research Center
P.O. Box 12235
Research Triangle Park,
North Carolina 27709
Phone (919) 834-7357 or 549-8291

Western Research Applications Center
University of Southern California
Los Angeles, California 90007
Phone (213) 746-6133

Anyone wishing to consult NASA concerning the availability of publications and documents may do so at a large number of public, university, and other libraries. The most nearly complete collections are available at the public libraries (usually the central ones) in the following cities: **California,** Los Angeles, San Diego; **Colorado,** Denver; **Connecticut,** Hartford; **Delaware,** Wilmington (Wilmington Institute Free Library); **Maryland,** Baltimore (Enoch Pratt Free Library); **Massachusetts,** Boston; **Michigan,** Detroit; **Minnesota,** St. Paul, Minneapolis; **Missouri,** Kansas City, St. Louis; **New Jersey,** Trenton; **New York,** New York, Brooklyn, Buffalo, Rochester; **Ohio,** Cleveland, Cincinnati, Dayton, Toledo, Akron; **Oklahoma,** Oklahoma City; **Tennessee,** Memphis; **Texas,** Fort Worth, Dallas; **Washington,** Seattle; **Wisconsin,** Milwaukee.

In addition, NASA's technical documents and bibliographic tools are desposited in 11 special libraries. Each library listed below is prepared to furnish to the general public services of personal reference, interlibrary loans, photocopies, and help in obtaining personal copies of NASA documents on microfiche if requested. These special libraries are located as follows: **California,** University of California Library, Berkeley; **Colorado,** University of Colorado Libraries, Boulder; **District of Columbia,** Library of Congress; **Georgia,** Georgia Institute of Technology, Atlanta; **Illinois,** The John Crerar Library, Chicago; **Massachusetts,** Massachusetts Institute of Technology, Cambridge; **Missouri,** Linda Hall Library, Kansas City; **New York,** Columbia University, New York; **Pennsylvania,** Carnegie Library, Pittsburgh; **Texas,** Southern Methodist University, Dallas; **Washington,** University of Washington Library, Seattle.

Additional information may be obtained by writing to: **Director, Technology Utilization Division, Code UT, NASA, Washington, D.C. 20546.**

There can be no question that the formulation of more efficient methods of applying advanced technology, as epitomized by the space program, is a desirable goal. Because the programs outlined above are still relatively new, it is understandable that inefficiencies exist; and the very multitude of documents and variety of sources often confuse the potential user and even compound his research problems. The development of improved communications and information systems presents us with one of the most challenging goals of all in applying science and technology to our needs.

SPECIAL PUBLICATIONS FOR EXPANDING OUR KNOWLEDGE

One of the principal benefits of the NASA technology utilization program has been the coordination and broad distribution of Special Publications (SP), reference manuals, handbooks, conference proceedings, and bibliographies which organize and summarize technical information in readily usable packages. These cover a wide scope of subjects and disciplines and have been used by numerous groups and individuals within NASA and by other government agencies, industry, and educational institutions. Selected examples are listed below, with the NASA document number in case copies are desired.

Applications for Systems Analysis Models (SP-5048)

This survey describes management models and systems analysis technology developed for NASA programs, and their potential use in nonaerospace programs. Industrial applications might include long-range planning, optimization of complex production systems, cost control, market research, and information systems. In addition, transportation, urban and regional planning, city administration, education, public health programs, and other areas of public and social application are discussed in this document.

Design Review for Space Systems (SP-6502)

This publication serves as an aid for instructing technical personnel in the implementation and evaluation of a design review program. Independent evaluation of a product prior to the testing phase is necessary to achieve reliability and assure that the design satisfies program requirements. The elements of a design review program, including participants, milestone categories, cost, data inputs, and data outputs, are clearly described in this document.

Management Procedures for Automatic Data Processing Equipment (NASA TM-X-59262)

This manual discusses policies and procedures for management of automatic data processing resources. It includes the selection, acquisition, evaluation, and review of all automatic data processing equipment primarily related to electronic digital computers or punched-card electric accounting machines. Auxiliary equipment evaluation methods, computer system specifications, and reference documents are also covered.

Apollo Documentation Administration Manual (NASA TM-X-59409)

Procedures, methods, and practices of NASA are presented in this manual for effective management of program documentation systems. Procedures for identification, planning, selection, acquisition, control, and scheduling of essential documents of the Apollo program are established.

Approval of Facility Projects (NASA TM-X-59259)

Policies and procedures for facility management presented in this handbook are primarily applicable to NASA installations but are possibly adaptable for industrial facilities. Facility acquisition, design, construction, repair, and alteration are covered.

Industrial Safety Handbook (NASA TM-X-57607)

This Kennedy Space Center handbook describes the recommended safety devices, procedures, and methods applicable to industrial equipment and operations at the center. It supplements approved safety codes, design criteria, and working methods.

Hydrogen Safety Manual (Tech Brief 68-10323)

This publication covers the natural characteristics of hydrogen, design principles for hydrogen systems, protection of personnel and equipment, and operating and emergency procedures. It sets standards and practices for minimum safety requirements at hydrogen installations.

Radiological Safety Handbook (NASA TM-X-54859)

This handbook describes organization, policies, and procedures for a radiological safety program. It can be used as a safety supplement for regulations for the use of X-ray equipment and radioactive materials. Handling and disposal of radioactive materials, personnel monitoring, dosage rates, and emergency procedures are discussed.

Contamination Control Handbook (SP-5076)

This handbook provides technical information for avoiding contamination of physical, chemical, or biological systems or products. Control methods are included for product design, treatment of gases and liquids, airborne and surface contamination, radiation, packaging, handling, storage, and personnel. (Also see SP-5045, SP-108, and SP-5074.)

Training Manuals for Nondestructive Testing Using Magnetic Particles (Tech Brief 68-10391)

These training manuals contain the fundamentals of nondestructive testing using magnetic particles as a detection medium. They can be used by inspectors of metal parts and by quality assurance specialists. Magnetic particle testing involves magnetization of the test specimen, application of magnetic particles, and interpretation of the patterns formed. (Also see SP-5082 and Tech Brief 67-10374.)

Training Manual on Optical Alignment Instruments (Tech Brief 68-10574)

This manual provides a basic course of instruction in the use of optical instruments for precise dimensional control and alignment of structural elements and assemblies, such as associated with space vehicles, aircraft, ships, and buildings.

Handbook for Magnetic Phenomena (NASA CR-58591)

This publication discusses the origin of magnetic effects, magnetic shielding, high- and low-field measurement methods, and testing in a shielded enclosure. Detailed definitions of terms, formulas, conversion tables, and a glossary are included.

Electromagnetic Compatibility Principles and Practices (NASA TM-X-57183)

This manual was prepared to stimulate interest in and develop understanding of electromagnetic problems and control in the Apollo program. The scientific background of electromagnetic compatibility; elements of electromagnetic interference; characteristics of electronic parts, circuits, and systems; and grounding, bonding, shielding, and interfacing specifications and measurement techniques are covered.

Reliable Electrical Connections—Technology Handbook (SP-5002)

This technology utilization handbook provides diagrams, photographs, and detailed instructions for various techniques and types of reliable electrical connections. The data, based primarily on the highly successful Saturn rocket program developed by the Marshall Space Flight Center of NASA, have obvious industrial applications.

Microelectronics in Space Research (SP-5031)

The NASA contributions to the microelectronic field are summarized in this technology utilization publication. Techniques, design, development, reliability, and applications are emphasized, with particular attention to silicon-integrated devices.

Batteries for Space Power Systems (SP-172)

This survey collects and summarizes all significant NASA-sponsored work published on batteries and related electrochemistry from 1959 through 1966. Battery principles, facts of high- and low-temperature operation, and detailed data for nickel-cadmium, silver-cadmium, silver-zinc, and lead-acid batteries are given. (Also see SP-5004.)

Pyrometry Handbook (Tech Brief 66-10520)

The Lewis Research Center developed this handbook dealing with surface temperature measurement by optical and ratio pyrometers. The handbook, which contains reference literature, results from experiments to provide a collection of applied technology and reference sources for engineers and technicians concerned with problems of measuring the surface temperatures of opaque materials. With a pyrometer designed from the handbook, a Midwestern electronics manufacturer has been able to make more precise measurements of the temperature of graphite used in the manufacture of semiconductors.

Welding for Electronic Assemblies (SP-5011)

Space-related welding technology for potential industrial use is here presented in the following studies: "Fundamentals," "Determination of Optimum Parameters," "Equipment," "Interconnecting and Component Lead Materials," "Weld Inspection," and "Process Control." Also included in this technology utilization document is a glossary of welding terms. (For other welding data see SP-5024; for soldering electrical connections see SP-5002.)

Separable Connector Design Handbook (NASA CR-64944)

This technical document can be used as an aid in the design and selection of fluid connections. Flanged and threaded designs, seals, leakage measurement, materials, configurations, and design compromises are covered in this publication, which was developed under contract to the Marshall Space Flight Center. (Also see SP-5903 and SP-5905.)

Fluid Properties Handbook (Tech Brief 67-10440)

This single-source document compiles accurate data pertaining to physical properties of helium, hydrogen, oxygen, and nitrogen.

Information contained in the handbook, prepared for the Marshall Space Flight Center, has proved valuable to many companies that manufacture processing and cryogenic equipment and instrumentation. The handbook has been particularly helpful in equipment design (as for fluidically controlled machine tools), training of new personnel, and calculations of density, volume, viscosity, and other characteristics of fluids.

Handbook of Cryogenic Data in Graphic Form (Tech Brief 67-10610)

This handbook is written in graphic form and concentrates extensive data concerning common materials of construction and properties of fluids frequently encountered in designing cryogenic systems. All data are presented in the British system of units.

Compressed Gas Handbook (SP-3045)

This technical handbook presents a discussion of the properties of high-pressure gases and pneumatic systems with appropriate theory and empirical data. Design-oriented information concerning hardware and practical information and applications are also presented, including detailed appendixes of general engineering data on air, nitrogen, oxygen, helium, and hydrogen.

Vacuum Technology and Space Simulation (SP-105)

This comprehensive vacuum manual combines material on space vacuum simulation and technology, and space chamber pressure measurement in one handy reference. Pumping systems, leak detection, outgassing problems, molecular fundamentals, and terminology are included in the coverage.

BIBLIOGRAPHY

CHAPTER 1

Aerospace Related Technology for Industry, SP-5075. Langley, Va., 1969: Langley Research Center, National Aeronautics and Space Administration.

The American Assembly—Outer Space: Prospects for Man and Society. New York, 1968: Columbia University.

The Commercial Application of Missile/Space Technology. Denver, 1963: University of Denver.

Significant Achievements in Space Applications, SP-156. Washington, D.C., 1967: National Aeronautics and Space Administration.

Space Program Benefits. (Hearing before the Committee on Aeronautical and Space Sciences, U.S. Senate.) Washington, D.C., 1970.

A Survey of Space Applications, SP-142. Washington, D.C., 1967: National Aeronautics and Space Administration.

Transforming and Using Space-Research Knowledge, SP-5018. Washington, D.C., 1964: National Aeronautics and Space Administration.

Alexander, T., "The Unexpected Payoff," *Fortune*, Vol. 80, No. 1 (July 1969).

Caidin, M., *Why Space? And How It Serves You in Your Daily Life*. New York, 1965: Messner.

Haggerty, J. J., "The Giant Harvest from Space—Today and Tomorrow," *Air Force/Space Digest*, Vol. 53, No. 2 (February 1970).

Ruzic, N. P., "The Case for Technological Transfer," *Industrial Research*, March 1965.

CHAPTER 2

"Better Medical Care in the 70's Based on Engineer-Doctor Teams," *Product Engineering*, Jan. 1, 1970.

Medical and Biological Applications of Space Telemetry, SP-5023. Washington, D.C., 1965: National Aeronautics and Space Administration.

NASA Contributions to Bioinstrumentation Systems: A Survey, SP-5054. Washington, D.C., 1968: National Aeronautics and Space Administration.

"Radio-Controlled Nurse," *Industrial Research*, Vol. 9, No. 7 (June 1967).

Bean, M. A., and others, "Monitoring Patients Through Electronics," *American Journal of Nursing*, Vol. 63, No. 12 (April 1963).

Collen, M. F., "Periodic Health Examination Using an Automated Multitest Laboratory," *Journal of the American Medical Association*, Vol. 195, No. 10 (Mar. 7, 1967).

Englebardt, S. L., "Cryosurgery: Cold Approach to Hot Cases," *The Modern Hospital*, Vol. 108, No. 4 (April 1967).

Fishlock, D., *Man Modified: An Exploration of the Man-Machine Relationship*. London, 1969: Jonathan Cape.

Hartwig, C., and T. Bendersky, "Guidelines to the Application of Space Technology to Medicine," *Research/Development*, Vol. 17, No. 9 (September 1966).

Hendrickson, W. J., "Tradition Versus the Space Age," *American Journal of Psychiatry*, Vol. 124, No. 7 (January 1968).

Jacobson, B., "Endoradiosonde Techniques–A Survey," *Medical Electronics and Biological Engineering*, April–June 1963.

Ko, W. H., and L. E. Slater, "Bioengineering: A New Discipline," *Electronics*, June 14, 1965.

Loftas, T., "Biomedical Technology," *Science Journal*, Vol. 5, No. 6 (June 1969).

Longmore, D., *Machines in Medicine*. New York, 1969: Doubleday.

MacGuire, H., "Atomedic: Hospital in the Round Provides New Concept in Medical Care," *Internal Surgery Bulletin*, Vol. 45, No. 6 (June 1966).

———, "Atomedics: New Name for Space Age Medicine," *The Canadian Doctor*, Vol. 33, No. 1 (January 1967).

Mecklin, J. M., "Hospitals Need Management Even More than Money," *Fortune*, Vol. 81, No. 1 (January 1970).

Ordway, F. I., J. P. Gardner, and M. R. Sharpe, *Basic Astronautics: An Introduction to Space Science, Engineering, and Medicine*. Englewood Cliffs, N.J., 1962: Prentice-Hall.

Pelligra, R., and others, "Centrifuge as a Therapeutic Device," *Aerospace Medicine*, Vol. 41, No. 4 (April 1970).

Schwichtenberg, A. H., "Clinical Application of Space Medicine," *Journal of the American Medical Association*, Vol. 201, No. 4 (July 24, 1967).
Sharpe, M. R., *Living in Space: The Astronaut and His Environment*. New York, 1969: Doubleday.
Shepherd, E. C., "Research Help on Tap," *New Scientist*, Vol. 34, No. 549 (June 15, 1967).

CHAPTER 3

Applications of Systems Analysis: A Survey, SP-5048. Washington, D.C., 1968: National Aeronautics and Space Administration.
"Police Turn to Space Technology for Help in War on Crime," *Product Engineering*, Jan. 1, 1970.
"$6 Million Hospital Saving Claimed for Systems Method," *Technology Week*, Vol. 20, No. 26 (June 26, 1967).
"Systems Approach: Political Interest Arises," *Science*, Vol. 153, No. 3741 (Sept. 9, 1966).

Alexander, T., "Shipbuilding's Big Lift from Aerospace," *Fortune*, Vol. 78, No. 3 (Sept. 1, 1968).
Bulban, E. J., "Automated Baggage System Tested for Use by Airlines," *Aviation Week & Space Technology*, Vol. 91, No. 16 (Oct. 20, 1969).
Flagle, C. D., "Operations Research and Health Services," *Operations Research*, Vol. 10, No. 5 (September–October 1962).
Gregory, W. H., "Industry Probes Socio-Economic Markets," *Aviation Week & Space Technology*, Vol. 88, No. 24 (June 10, 1968).
Herman, C. C., "Systems Approach to City Planning," *Harvard Business Review*, Vol. 44, No. 5 (September–October 1966).
Katz, T., "Philosophy and Education in the Space Age," *New York Academy of Science Annals*, Vol. 140 (1965).
Kircher, D. P., "Corporate Participation in the Solution of Social Problems," *Astronautics & Aeronautics*, Vol. 7, No. 10 (October 1969).
Lessing, L., "Systems Engineering Invades the City," *Fortune*, Vol. 77, No. 1 (January 1968).
Oettinger, A. G., "Myths of Educational Technology," *Saturday Review*, Vol. 51, No. 25 (May 18, 1969).
Ramo, S., *Cure for Chaos: Fresh Solutions to Social Problems Through the Systems Approach*. New York, 1969: McKay.
———, "The Systems Approach: Automated Common Sense," *Nation's Cities*. Washington, D.C., March 1968: National League of Cities.
Ross, H. R., "New Transportation Technology," *International Science and Technology*, No. 59 (November 1966).
Sharpe, M. R., *Living in Space: The Astronaut and His Environment*. New York, 1969: Doubleday.
Slomski, J. F., "Systems Engineering Management: Its Principles and Practice," *Aerospace Management*, Vol. 4, No. 2 (February 1969).

Wildavsky, A., "The Political Economy of Efficiency: Cost-Benefit Analysis, Systems Analysis, and Program Budgeting," *Public Administration Review*, Vol. 24, No. 4 (December 1966).
Wilks, W. E., "Industry Studies California Problems," *Missiles and Rockets*, Vol. 16, No. 7 (Feb. 15, 1965).

CHAPTERS 4, 5, 6, 7

These four chapters involve various aspects of orbital observation and remote sensing of the Earth's ocean and land surfaces and of the atmosphere. Because of the fact that so many titles deal with more than a single subject, it was found to be more practical to prepare a single bibliography rather than separate bibliographies for each of the four chapters. This approach not only avoids redundant listings but helps to lead the reader to multisubject articles, reports, and books.

Limitations of available space make it impossible to include more than a relatively small portion of the literature studied during the course of preparation of these chapters. The reader is encouraged to acquire an excellent survey, published in September 1970 by the National Aeronautics and Space Administration, entitled *Remote Sensing of Earth Resources: A Literature Survey with Indexes* (SP-7036). It has 665 pages plus subject, author, and other indexes. The subject categories are as follows:

Agriculture and Forestry; Environmental Changes and Cultural Resources; Geodesy and Cartography; Geology and Mineral Resources; Oceanography and Marine Resources; Hydrology and Water Management; Data Processing and Distribution Systems; Instrumentation and Sensors; and General.

In the remote-sensing field, the reader can consult the quarterly interdisciplinary journal *Remote Sensing of Environment* and the *Journal of Remote Sensing*, published by the International Remote Sensing Institute, as well as the more generally known periodical literature covering the Earth-resources and remote-sensing fields, for example, *Photogrammetric Engineering*. Because of the newly expanded interest in oceanography, some of the journals and magazines covering this field are listed below:

Marine Geophysical Researches, D. Riedel Publishing Co.
Ocean Engineering, Pergamon Press (monthly)
Oceanic Citation Journal, Oceanic Research Institute (monthly)
Oceanic Index, Oceanic Library and Information Center, La Jolla, Calif.
Ocean Industry, Gulf Publishing Co. (monthly)
Oceanology, Ziff-Davis Publishing Co. (weekly)
Oceanology International, Industrial Research (seven times a year)
Oceans, Oceans Publishers, Inc. (monthly)
Ocean Science News, Nautilus Press, Inc. (weekly)
Undersea Technology, Compass Publications (monthly)

Underwater Science and Technology Bulletin (monthly)
Underwater Science and Technology Journal (quarterly)

Apollo 6 and 7 Synoptic Photography Catalog and Apollo 9 Synoptic Photography Catalog. Albuquerque, 1968 and 1969: Technology Application Center, University of New Mexico.

Appraisal of the Potential for Agricultural Resource Surveys by Remote Sensing, CR-103704. Washington, D.C., 1969: Natural Resource Economics Division, U.S. Department of Agriculture.

Bibliography of Remote Sensing of Resources. Fort Belvoir, Va., 1966: U.S. Army Corps of Engineers.

Charts Designate Probable Future Oceanographic Research Fields, Tech Brief 68-10397. Huntsville, Ala., 1968: Marshall Space Flight Center, National Aeronautics and Space Administration.

Congressional Reports, 1969, 1970, and 1971 NASA authorizations. (Hearings before the Subcommittee on Space Science and Applications, Committee on Science and Astronautics, U.S. House of Representatives.) Washington, D.C., 1968, 1969, and 1970.

Earth Photographs from Gemini III, IV, and V and *Earth Photographs from Gemini VI Through XII,* SP-129 and SP-171. Washington, D.C., 1967 and 1968: National Aeronautics and Space Administration.

Earth Resources: Cooperative Research in Remote Sensing for Earth Surveys—Agreement Between United States and Brazil and *Earth Resources: Cooperative Research in Remote Sensing for Earth Surveys—Agreement Between United States and Mexico,* Publications 6569 and 6613. Washington, D.C., 1968: U.S. Department of State.

"Earth Resources Observation Satellite: An Overview of the World We Live on," *Environmental Science and Technology,* Vol. 1, No. 6 (June 1967).

Earth Resources Satellite System. (Report of the Subcommittee on NASA Oversight, Committee on Science and Astronautics, U.S. House of Representatives, Serial W.) Washington, D.C., 1969.

Ecological Surveys from Space, SP-230. Washington, D.C., 1970: National Aeronautics and Space Administration.

Effective Use of the Sea. Washington, D.C., 1966: President's Science Advisory Committee.

Exploring Space with a Camera, SP-138. Washington, D.C., 1968: National Aeronautics and Space Administration.

Feasibility of a Global Observation and Analysis Experiment, Publication 1290. Washington, D.C., 1966: National Research Council.

For the Benefit of All Mankind—a Survey of the Practical Returns from Space Investment. (Report of the Committee on Science and Astronautics, U.S. House of Representatives, Serial R.) Washington, D.C., 1970.

"Infrared Magic," *Agricultural Research,* Vol. 18, No. 1 (July 1969).

Man's Geophysical Environment: Its Study from Space. Washington, D.C., 1968: U.S. Government Printing Office.

Marine Science Affairs. Washington, D.C., 1967: National Council on Marine Resources and Engineering Development.

Meteorological Satellites and Sounding Rockets, EP-27. Washington, D.C., 1965: National Aeronautics and Space Administration.

New Horizons in Color Aerial Photography: Proceedings of the Seminar. Falls Church, Va., 1969: American Society of Photogrammetry.

Objectives and Goals in Space Science and Applications—1968, SP-162. Washington, D.C., 1968: National Aeronautics and Space Administration.

"The Ocean" (special issue), *Scientific American,* Vol. 221, No. 3 (September 1969).

"The Ocean and Marine Systems" (special issue), *Astronautics & Aeronautics,* Vol. 4, No. 4 (April 1966).

Oceanography 1966. (Report of the NASA Committee on Oceanography.) Washington, D.C., 1967: National Academy of Sciences.

Peaceful Uses of Earth-Observation Spacecraft, CR-587 and CR-588. Ann Arbor, Mich., and Washington, D.C., 1966: University of Michigan and National Aeronautics and Space Administration.

The Post-Apollo Space Program: Directions for the Future. (Report to the President.) Washington, D.C., 1969: Space Task Group.

Potential Applications of Satellite Geodetic Techniques to Geosciences, SP-158. Washington, D.C., 1968: National Aeronautics and Space Administration.

The Potential of Observation of the Oceans from Spacecraft, PB-177726. (General Electric Co. report for National Council on Marine Resources and Engineering Development.) Washington, D.C., 1967: Clearinghouse for Federal Scientific and Technical Information.

Proceedings of Symposia on Remote Sensing of the Environment. (6 symposia.) Ann Arbor, Mich., through 1969: Willow Run Laboratories, and Institute of Science and Technology, University of Michigan.

Proceedings of the Winter Study on the Uses of Manned Space Flight, 1975–1985, SP-196. Washington, D.C., 1969: National Aeronautics and Space Administration.

"Project TIREC," *American Meteorological Society Bulletin,* Vol. 43, No. 7 (1962).

"Results of Space Research: Meteorology," *TRW Space Log,* Vol. 4, No. 4 (Winter 1964–65).

Satellite Applications to Marine Geodesy, CR-1253. Columbus, 1969: Battelle.

Satellite Data in Meteorological Research, NCAR-TN-11. Boulder, Colo., 1966: National Center for Atmospheric Research.

Satellite Triangulation in the Coast and Geodetic Survey, Tech. Bull. 24. Washington, D.C., 1965: U.S. Commerce Department.

"Sensors Pose Earth Satellite Challenge," *Aviation Week & Space Technology,* Vol. 92, No. 25 (June 22, 1970).

Significant Achievements in Satellite Geodesy, 1958–1964, SP-94. Washington, D.C., 1966: National Aeronautics and Space Administration.

Significant Achievements in Satellite Meteorology, 1958–1964, SP-96. Washington, D.C., 1966: National Aeronautics and Space Administration.

Soviet Scientists Present Papers on Satellite Meteorology, JPRS-39599. Washington, D.C., 1967: Joint Publications Research Service.

Spacecraft in Geographic Research. Washington, D.C., 1966: National Academy of Sciences.

Spacecraft Oceanography Project Review of NASA Earth Resources Aircraft Program and Remote Sensing Experiments. Washington, D.C., 1968: U.S. Naval Oceanographic Office.

Space Program Benefits. (Hearing before the Committee on Aeronautical and Space Sciences, U.S. Senate.) Washington, D.C., 1970.

SPOC–Spacecraft Oceanography Project. (Annual reports.) Washington, D.C., 1960–1970: U.S. Naval Oceanographic Office.

A Survey of Space Applications, SP-142. Washington, D.C., 1967: National Aeronautics and Space Administration.

TRIAD: Preliminary Design of an Earth Resources Survey System, CR-106276. Norfolk, Va., 1969: Old Dominion College.

United Nations Conference on the Exploration and Peaceful Uses of Outer Space. Vienna, August 1968.

United States Activities in Spacecraft Oceanography. Washington, D.C., 1967: U.S. Government Printing Office.

Useful Applications of Earth-Oriented Satellites. Washington, D.C., 1969: National Academy of Sciences.

World Weather Program. Washington, D.C., 1970: U.S. Government Printing Office.

Adler, C., "Ocean Pollution Problems," *Science and Technology*, No. 93 (October 1969).

Albanese, D. F., "Geodetic Satellite," *Electrical Communications*, Vol. 43, No. 1 (1968).

Anderson, R. E., "Satellite Navigation and Communication for Merchant Ships," *Navigation*, Vol. 14, No. 2 (Summer 1967).

Arnold, E., "Sea Surface Temperature Anomalies and Oceanic Cloud Distribution," in *Ocean from Space Symposium.* Houston, 1967: American Society for Oceanography.

Arnold, K., "The Use of Satellites for Geodetic Studies," *Space Science Reviews*, Vol. 7, No. 1 (August 1967).

Badgley, P. C. (ed.), *Scientific Experiments for Manned Orbital Flight*, Science and Technology Series, Vol. 4. North Hollywood, Calif., 1965: Western Periodicals.

———, and others, "NASA Earth-Sensing Space Flight Experiments," *Photogrammetric Engineering*, Vol. 34, No. 2 (February 1968).

———, L. Miloy, and L. F. Childs, *Oceans from Space.* Houston, 1970: Gulf Publishing.

———, and W. L. Vest, "Orbital Remote Sensing and Natural Resources," *Photogrammetric Engineering*, Vol. 32, No. 5 (September 1966).

Bailey, J. S., and P. G. White, "Remote Sensing of Ocean Color," in *Advances in Instrumentation*, Vol. 24, Part 3. Pittsburgh, 1969: Instrument Society of America.

Barnes, H. (ed.), *Oceanography and Marine Biology*. (5 vols.) London, 1963–67: Allen & Unwin.

———, D. T. Chang, and J. H. Willand, *Satellite Infrared Observation of Arctic Sea Ice*, Paper 70-301. New York, 1970: American Institute of Aeronautics and Astronautics.

Barringer, A. R., "Detecting the Ocean's Food (and Pollutants) from Space," *Ocean Industry*, May 1967.

———, J. H. Davies, and A. J. Moffat, *The Problems and Potential in Monitoring Pollution from Satellites*, Paper 70-305. New York, 1970: American Institute of Aeronautics and Astronautics.

Behrman, D., *Exploring the Ocean World*. Boston, 1970: Little, Brown.

Bird, J. B., and A. Morrison, "Space Photography and Its Geographical Applications," *Geographical Review*, Vol. 54, No. 4 (October 1964).

Bylinsky, G., "From a High-Flying Technology–a Fresh View of Earth," *Fortune*, Vol. 78, No. 7 (June 1, 1968).

Campbell, S. G., "Global Meteorology: A Systems Approach," *Bulletin of the American Meteorological Society*, Vol. 45, No. 1 (January 1964).

Canney, H. E., and F. I. Ordway, "The Uses of Artificial Satellite Vehicles," *Astronautica Acta*, Part 2, Fasc. 4 (1956) and Part 3, Fasc. 1 (1957).

Carter, L. J., "Earth Resources Satellite: Finally off the Ground?" *Science*, Vol. 163, No. 3869 (Feb. 21, 1969).

Chapman, W. M., "How Space Research Can Help Develop Fisheries," *Ocean Industries*, Vol. 2, No. 5 (1967).

Colvocoresses, A. P., "Surveying the Earth from 20,000 Miles," in *Proceedings of the Annual Convention of Photographic Scientists and Engineers –Image Technology*. Washington, D.C., 1969: Acolyte.

Colwell, R. N., "Remote Sensing of Natural Resources," *Scientific American*, Vol. 218, No. 1 (January 1968).

Crutchfield, J. A., and others, *Observations from Satellites: Potential Impact on the United States Fishery, Some Considerations Affecting Satellite Data Contributions to Fisheries, Sea Surface Effects and Considerations Related to Remote Sensing from Orbiting Satellites, State of the Art, All Weather Sea Surface Sensing*, N00-SPOC-TN-1-4. Palo Alto, Calif., 1967: Space and Re-Entry Systems Division, Philco-Ford Corp.

Curnow, R. C., "The Importance of Application Satellites to the European Economy," *Aeronautical Journal*, Vol. 73, No. 4 (April 1969).

Draeger, W. C., *The Interpretability of High Altitude Multispectral Imagery for the Evaluation of Wildland Resources*, CR-97824. Berkeley,

Calif., 1968: School of Forestry and Conservation, University of California.

———, and D. T. Lauer, *Present and Future Forestry Applications of Remote Sensing from Space*, Paper 67-765. New York, 1967: American Institute of Aeronautics and Astronautics.

Dulberger, L. H., "Geodetic Measurement from Space," *Space/Aeronautics*, Vol. 43, No. 6 (June 1966).

Emery, K. O., and E. Sinham, *Oceanographic Books of the World, 1957–1966*. Washington, D.C., 1967: Marine Technology Society.

Enzmann, R. D. (ed.), *Use of Space Systems for Planetary Geology and Geophysics*, Science and Technology Series, Vol. 17. Tarzana, Calif., 1968: American Astronautical Society Publications Office.

Ewing, G. C., *Oceanography from Space—Proceedings of Conference on the Feasibility of Conducting Oceanographic Explorations from Aircraft, Manned Orbital and Lunar Laboratories*. Woods Hole, Mass., 1965: Woods Hole Oceanographic Institution.

Fischer, W. A., "EROS: Investigations from Space," *Ground Water Resources Institute Quarterly*, Vol. 2, No. 1 (Spring 1969).

Forbes, L., and M. Sears, *Oceanography in Print*. Falmouth, Mass., 1968: Sailing Book Service (in cooperation with Woods Hole Oceanographic Institution, U.S. Bureau of Commercial Fisheries, and Marine Biological Laboratory).

Ford, C. Q. (ed.), *Space Technology and Earth Problems*, Science and Technology Series, Vol. 23. Tarzana, Calif., 1970: American Astronautical Society Publications Office.

Frey, H. T., *Agricultural Application of Remote Sensing: The Potential from Space Platforms*, Info. Bull. 328. Washington, D.C., 1967: Natural Resource Economics Division, U.S. Department of Agriculture.

Fritz, S., "Pictures from Meteorological Satellites and Their Interpretation," *Space Science Reviews*, Vol. 3, No. 4 (November 1964).

Gerlach, A. C., "The Geographical Applications Program of the U.S. Geological Survey," *Photogrammetric Engineering*, Vol. 35, No. 1 (January 1969).

Gilmer, J. R., A. M. Mayo, and R. C. Peavey (eds.), *Commercial Utilization of Space*, Advances in the Astronautical Sciences Series, Vol. 23. Tarzana, Calif., 1968: American Astronautical Society Publications Office.

Gordon, T. J., L. M. Dicke, and J. S. Nieroski, "Economics of Commercial Space Stations," in *Proceedings of the 18th International Astronautical Congress*, Vol. 2 (ed. by M. Lunc). Oxford, 1968: Pergamon.

Graham, L. C., "Earth Resources Determination with Terrain Imaging Radar," in R. Bundy (ed.), *Resources Roundup*. Phoenix, 1969: Institute of Electrical and Electronics Engineers.

Greenwood, J. A., and others, "Oceanographic Applications of Radar Altimetry from a Spacecraft," *Remote Sensing of Environment*, Vol. 1 (March 1969).

Guier, W. H., "Satellite Geodesy," *APL Technical Digest*, Vol. 4, No. 3 (January–February 1965).

Haggerty, J. J., "The Giant Harvest from Space–Today and Tomorrow," *Air Force/Space Digest*, Vol. 53, No. 2 (February 1970).

Hallgren, R. E., "The World Weather Program," *TRW Space Log*, Vol. 8, No. 1 (Spring 1968) and 2 (Summer).

Hanessian, J., and J. M. Logsdon, "Earth Resources Technology Satellite –Securing International Participation," *Astronautics & Aeronautics*, Vol. 8, No. 8 (August 1970).

Hardy, E. E., *Potential Benefits to Be Derived from Applications of Remote Sensing of Agricultural, Forest, and Range Resources*, Paper 69-583. Washington, D.C., 1969: American Astronautical Society.

Haviland, R. P., "On Applications of the Satellite Vehicle," *Jet Propulsion*, Vol. 26, No. 5 (May 1956).

Hemphill, W. R., and W. Danilchik, "Geologic Interpretation of a Gemini Photo," *Photogrammetric Engineering*, Vol. 34, No. 2 (February 1968).

Hieronymus, W. S., "Resource Satellite Effort Spurred," *Aviation Week & Space Technology*, Vol. 91, No. 20 (Nov. 17, 1969).

Idyll, C. P. (ed.), *Exploring the Ocean World: A History of Oceanography*, New York, 1969: Thomas Y. Crowell.

Jaffe, L., "Space Applications: Growing Worldwide Systems," *Astronautics & Aeronautics*, Vol. 4, No. 6 (June 1966).

———, and R. A. Summers, "The Earth Resources Survey Program Jells," *Astronautics & Aeronautics*, April 1971.

Johnson, A. W., "Weather Satellites," *Scientific American*, Vol. 220, No. 1 (January 1969).

Jurkevich, I., "Should Astronomical Observations Be Made from Manned Interplanetary Spacecraft?" *Astronautics & Aeronautics*, Vol. 8, No. 6 (June 1970).

Karth, J. E., *Earth Resources Surveys: An Outlook to the Future*. (Hearing before the Committee on Science and Astronautics, U.S. House of Representatives.) Washington, D.C., 1969.

Katz, A. H., "Observation Satellites: Problems and Prospects," *Astronautics*, Vol. 5, No. 4 (April 1960), No. 6 (June), No. 7 (July), No. 8 (August), No. 9 (September), and No. 10 (October).

Kaula, W. M., *Theory of Satellite Geodesy*. Waltham, Mass., 1966: Blaisdell.

Kavanau, L. L. (ed.), *Practical Space Applications*, Advances in the Astronautical Sciences Series, Vol. 21. Sun Valley, Calif., 1967: Scholarly Publications.

Krishna, R. P., and others, "Remote Sensing of Sea Surface Temperature," in *Proceedings of the Sixth Space Congress Canaveral Council of Technical Societies*, Vol. 2, *Space, Technology and Society*. Cape Canaveral, Fla., 1969: Canaveral Council of Technical Societies.

Lent, J. D., *Cloud Cover Interference with Remote Sensing of Forested Areas from Earth-Orbital and Lower Altitudes*, N66-39303. Berkeley, Calif., 1966: School of Forestry, University of California.

Lepley, L. K., "Coastal Water Clarity from Space Photographs," *Photogrammetric Engineering*, Vol. 34, No. 7 (July 1968).

Liapunov, B. V., *Station Outside the Earth*, FTD-MT-64-531. Wright-Patterson Air Force Base, Ohio, 1966: Foreign Technology Division, U.S. Air Force.

Lieberman, A., and P. Schipma, *Air-Pollution-Monitoring Instrumentation*, SP-5072. Washington, D.C., 1969: National Aeronautics and Space Administration.

Liventsov, A. V., *Sensitive Elements and Sensors Used on Spacecraft for Scientific Missions, USSR.*, Report 50113. Washington, D.C., 1970: Joint Publications Research Service.

Lowman, P. D., *Geologic Applications of Orbital Photography*, TN D-4155. Greenbelt, Md., 1967: Goddard Space Flight Center, National Aeronautics and Space Administration.

———, "Photography from Space," *Science Journal*, Vol. 1, No. 3 (May 1965).

———, *A Review of Photography of the Earth from Sounding Rockets and Satellites*, TN D-1868. Washington, D.C., 1964: National Aeronautics and Space Administration.

———, "Space Photography—A Review," *Photogrammetric Engineering*, Vol. 31, No. 1 (January 1965).

Ludwick, E. E., "Space Oceanography—Applications and Benefits," in *Proceedings of the Sixth Space Congress Canaveral Council of Technical Societies*, Vol. 2, *Space, Technology and Society*. Cape Canaveral, Fla., 1969: Canaveral Council of Technical Societies.

Ludwig, E. B., R. Bartle, and M. Griggs, *Study of Air Pollutant Detection by Remote Sensors*, CR-1380. Washington, D.C., 1969: National Aeronautics and Space Administration.

McClain, E. P., *Potential Use of Earth Satellites for Solving Problems in Oceanography and Hydrology*, Paper 69-596. Washington, D.C., 1969: American Astronautical Society.

MacKallor, J. A., *A Photomosaic of Western Peru from Gemini Photography*, Paper 600-C. Washington, D.C., 1968: U.S. Geological Survey.

Mannella, G. G., "Aerospace Sensor Systems," *Astronautics & Aeronautics*, Vol. 6, No. 12 (December 1968).

Markham, W. E., and R. W. Popham, *Operational Satellite Ice Reconnaissance and Surveillance, Tiros V and VI.* (Joint reports of Canadian Department of Transport and U.S. Weather Bureau.) October 1963.

Maughan, P. M., "Remote-Sensor Applications in Fishery Research," *Marine Technology Society Journal*, Vol. 3, No. 2 (March 1969).

Mekel, J. F. M., *Geology from the Air*. Delft, Netherlands, 1969: Uitgevezij Waltman.

Merifield, P. M., and J. Rammelkamp, "Terrain Seen from Tiros," *Photogrammetric Engineering*, Vol. 32, No. 1 (January 1966).

Moody, J. C., and O. Weinstein, "Night and Day Nimbus 2 Transmits Its Cloud Pictures," *Electronics*, Vol. 39, No. 17 (Aug. 22, 1966).

Moore, E., and B. S. Wellar, "Experimental Applications of Multiband Photography in Urban Research," *Transactions of Illinois State Academy of Science*, Vol. 61, No. 1 (1968).

Moore, R. K., and D. S. Simonett, "Radar Remote Sensing in Biology," *Bioscience*, Vol. 17, No. 6 (June 1967).

Morrison, A., and M. C. Chown, *Photography of the Western Sahara from the Mercury MA-4 Spacecraft*, CR-126. Washington, D.C., 1964: National Aeronautics and Space Administration.

Mueller, G. E., "Earthly Dividends from Space," *Spaceflight*, Vol. 11, No. 12 (December 1969).

Mueller, I. I., *Introduction to Satellite Geodesy*. New York, 1964: Ungar.

Narin, F. (ed.), *Post Apollo Space Exploration*, Advances in the Astronautical Sciences Series, Vol. 20. Tarzana, Calif., 1966: American Astronautical Society Publications Office.

Neumann, G., and W. J. Pierson, *Principles of Physical Oceanography*. Englewood Cliffs, N.J., 1968: Prentice-Hall.

Newell, H. E., "Current Programme and Considerations of the Future Earth Resources Survey," *Spaceflight*, Vol. 10, No. 8 (August 1968).

Newhall, B., *Airborne Camera: The World from the Air and Outer Space*. New York, 1969: Hastings.

Newton, R. R., "Geodesy by Satellite," *Science*, Vol. 144, No. 3620 (May 15, 1964).

Nicolaides, J. D., M. M. Macomber, and W. M. Kaula, "Terrestrial, Lunar and Planetary Applications of Navigation and Geodetic Satellites," in *Advances in Space Science and Technology*, Vol. 5 (ed. by F. I. Ordway). New York, 1963: Academic.

Norton, V. J., *Some Potential Benefits to Commercial Fishing Through Increased Search Efficiency: A Case Study—The Tuna Industry*. Kingston, R.I., 1969: Department of Food and Resource Economics, University of Rhode Island.

Olson, B. E., *Remote Sensing in Oceanography*, IR 68-18. Washington, D.C., 1968: Naval Oceanographic Office.

Ordway, F. I., "Think About This!—Geology Has Great Potential in the Space Age," *Geotimes*, Vol. 2, No. 10 (April 1958).

Ostrow, H., and O. Weinstein, "A Review of a Decade of Space Camera Systems Development for Meteorology," in *Proceedings, Society of Photo-Optical Instrumentation Engineers 13th Annual Technical Symposium*, Vol. 1. Redondo Beach, Calif., 1969: Society of Photo-Optical Instrumentation Engineers.

Pardoe, G. K. C., "Earth Resource Satellites," *Science Journal*, Vol. 5, No. 6 (June 1969).

Park, A. B., *What Earth Resources Satellites Can Do for the Agricultural Community*, Paper 69-1083. New York, 1969: American Institute of Aeronautics and Astronautics.

Pecora, W. T., "Surveying the Earth's Resources from Space," *Surveying and Mapping*, Vol. 27, No. 4 (April 1967).

Pesce, A., *Gemini Space Photographs of Libya and Tibesti.* Tripoli, 1968: Petroleum Exploration Society of Libya.

Polezhayev, A. P., *Use of Artificial Earth Satellites for Geodesy*, JPRS-38537. Washington, D.C., 1966: Joint Publications Research Service.

Pouquet, J., *An Approach to the Remote Detection of Earth Resources in Sub-Arid Lands,* TN-D-4647. Greenbelt, Md., 1968: Goddard Space Flight Center, National Aeronautics and Space Administration.

Powell, W. J., C. W. Copeland, and J. A. Drahovzal, *Delineation of Linear Features and Application to Reservoir Engineering Using Apollo 9 Multispectral Photography*, Report In. Series 41. February 1970: Geological Survey of Alabama, University of Alabama.

Press, H., and W. B. Huston, "Nimbus: A Progress Report," *Astronautics & Aeronautics,* Vol. 6, No. 3 (March 1968).

Robinove, C. J., *Future Applications of Earth Resource Surveys from Space,* Paper 70-302. New York, 1970: American Institute of Aeronautics and Astronautics.

———, "Perception via Satellite," *Water Spectrum,* Vol. 2, No. 1 (Spring 1970).

———, "Remote-Sensing Potential in Basic-Data Acquisition," in *Proceedings of the International Conference on Water for Peace,* Washington, D.C., 1967.

———, and D. G. Anderson, "Some Guidelines for Remote Sensing in Hydrology," *Bulletin of the American Water Resources Association* (1969).

Rouse, J. W., *Ice Type Identification by Radar.* Lawrence, Kans., 1968: Center for Research, University of Kansas.

———, and others, "Use of Orbital Radars for Geoscience Investigations," in *Proceedings of the Third Space Congress.* Cocoa Beach, Fla., 1966: Canaveral Council of Technical Societies.

Ruppe, H. O., "Astronautics: An Outline of Utility," in *Advances in Space Science and Technology,* Vol. 10 (ed. by F. I. Ordway). New York, 1970: Academic.

Salomonson, V. V., *Cloud Statistics in Earth Resources Technology Satellite (ERTS) Mission Planning,* TM-X-63674. Greenbelt, Md., 1969: Goddard Space Flight Center, National Aeronautics and Space Administration.

Saunders, P. M., "The Temperature of the Ocean-Air Interface," *Journal of the Atmospheric Sciences,* Vol. 24, No. 3 (May 1967).

Schertler, R. J., *Remote Sensing of Agriculture, Forestry and Water Resources.* Chicago, 1969: American Society of Farm Managers and Rural Appraisers Conference.

Schneider, W. J., "Color Photography for Water Resources Studies," *Photogrammetric Engineering,* Vol. 34, No. 3 (March 1968).

Sears, M. (ed.), *Oceanography.* Washington, D.C., 1961 and 1969: American Association for the Advancement of Science.

Shapiro, S., "Manned Space Stations–How and Why." *Science Journal,* Vol. 5, No. 2 (February 1969).

Sharpe, M. R., *Satellites and Probes: The Development of Unmanned Spaceflight.* New York, 1970: Doubleday.

Shifrin, K. S., and V. L. Gayevskiy (eds.), *Satellite Meteorology*, TT F-589. Washington, D.C., 1970: National Aeronautics and Space Administration.

Singer, S. F., "Satellite Meteorology," *International Science and Technology*, No. 36 (December 1964).

Slater, L. E., *Some Prospects of Using Communication Satellites in Wild Animal Research*, DDC AD 647 291. Washington, D.C., 1966: Bioinstrumentation Advisory Council, American Institute of Biological Sciences.

Smith, J. T. (ed.), *Manual of Color Aerial Photography*. New York, 1968: American Society of Photogrammetry.

Spindell, W. A., "Remote Sensing of Earth Resources Using Manned Spacecraft," in *Proceedings of the Sixth Space Congress*. Cocoa Beach, Fla., 1969: Canaveral Council of Technical Societies.

Stehling, K. R., "Remote Sensing of the Oceans," *Astronautics & Aeronautics*, Vol. 7, No. 5 (May 1969).

———, "Spotting Pollution from Space," *Space Aeronautics*, Vol. 53, No. 6 (June 1970).

Stevenson, R. E., *Oceanographic Applications and Limitations of Satellite Remote Sensors*, Paper 70-306. New York, 1970: American Institute of Aeronautics and Astronautics.

———, *A Real-Time Fisheries Satellite System*, Contribution 287. Galveston, Tex., 1969: Biological Laboratory, U.S. Bureau of Commercial Fisheries.

Stoertz, G. E., W. R. Hemphill, and D. A. Markle, "Airborne Fluorometer Applicable to Marine and Estuarine Studies," *Marine Technology Society Journal*, Vol. 3, No. 6 (November–December 1969).

Thomas, P. G., "Earth-Resource Survey from Space," *Space/Aeronautics*, Vol. 50, No. 1 (July 1968).

———, "Global Weather Forecasting," *Space/Aeronautics*, Vol. 48, No. 5 (October 1967).

Vaeth, J. G., "Establishing an Operational Weather Satellite System," in *Advances in Space Science and Technology*, Vol. 7 (ed. by F. I. Ordway). New York, 1965: Academic.

Vetlov, I., *The Soviet "Meteor" Space System* (translation). Washington, D.C., 1967: Joint Publications Research Service.

Von Braun, W., and F. I. Ordway, *History of Rocketry & Space Travel*. New York, 1969: Thomas Y. Crowell.

Wallace, R. E., and D. B. Slemmons, "Possible Applications for Remote-Sensing Techniques and Satellite Communications for Earthquake Studies," in *Proceedings of the United States–Japan Conference on Research Related to Earthquake Prediction*. Washington, D.C., 1966: National Academy of Sciences.

Warnecke, G., L. J. Allison, and L. L. Forshee, *Observations of Sea Surface Temperatures and Ocean Currents from Nimbus II*, X-622-67-435.

Greenbelt, Md., 1967: Goddard Space Flight Center, National Aeronautics and Space Administration.

———, L. M. McMillin, and L. J. Allison, *Ocean Current and Sea Surface Temperature Observations from Meteorological Satellites*, TN D-5142. Washington, D.C., 1969: National Aeronautics and Space Administration.

Warren, H. R., and others, "Design Considerations for Ceres: A Satellite to Survey Canada's Natural Resources," *Canadian Aeronautics and Space Journal*, Vol. 14, No. 4 (April 1968).

Weiss, M., "Application of Infrared Measuring Techniques to the Remote Sensing of Water Surface Temperature," *Canadian Aeronautics and Space Journal*, Vol. 15, No. 5 (May 1969).

———, "Remote Temperature Sensing," *Oceanology International*, Vol. 3, No. 6 (September–October 1968).

Widger, W. K., *Meteorological Satellites*. New York, 1966: Holt, Rinehart & Winston.

Wilkerson, J. C., "The Gulf Stream from Space," *Oceanus*, Vol. 3, Nos. 2 and 3 (June 1967).

Williams, O. W., "Surveying the Earth by Satellite," *Science Journal*, Vol. 3, No. 1 (January 1967).

Wilson, R. C., "Space Photography for Forestry," *Photogrammetric Engineering*, Vol. 33, No. 5 (May 1967).

Wujek, J. H., "Weather Surveillance by Satellite," *Electronics World*, Vol. 77, No. 3 (March 1967).

CHAPTER 8

"Global Comsat Net Proposed by Soviets as Intelsat Rival," *Aviation Week & Space Technology*, Vol. 89, No. 8 (Aug. 19, 1968).

"Suddenly Everybody Wants a Comsat, but Who Will Operate What?" *Space/Aeronautics*, Vol. 45, No. 5 (May 1966).

"TV Broadcast from Space: Direct or Distributed?" *Space/Aeronautics*, Vol. 50, No. 6 (November 1968).

Bradley, W. E., "Communications Strategy of Geostationary Orbit," *Astronautics & Aeronautics*, Vol. 6, No. 4 (April 1968).

Brown, R., "A Ceiling on Satellite Communications," *New Scientist*, Vol. 38, No. 595 (May 2, 1968).

Clarke, A., "Extra Terrestrial Relays," *Wireless World*, Vol. 51, No. 10 (October 1945).

Jaffe, L., *Communications in Space*. New York, 1966: Holt, Rinehart & Winston.

Jamison, D., and others, "Satellite Radio: Better than ETV," *Astronautics & Aeronautics*, Vol. 7, No. 10 (October 1969).

Kolcum, E. H., "French Plan Commercial Satellite System," *Aviation Week & Space Technology*, Vol. 91, No. 21 (Nov. 24, 1969).

Lessing, L., "Cinderella in the Sky: Satellite to Home Broadcasting," *Fortune*, Vol. 76, No. 10 (October 1967).
Mama, H. P., "Making Space Work for India," *Spaceflight*, Vol. 12, No. 2 (February 1970).
Meckling, W., "Economical Potential of Communications Satellites," *Science*, Vol. 133, No. 3648 (June 16, 1961).
Pierce, J. R., "Orbital Radio Relays," *Jet Propulsion*, Vol. 25, No. 4 (April 1955).
Pritchard, W. L., "Communications Satellites and the World of Tomorrow," *Astronautics &Aeronautics*, Vol. 6, No. 4 (April 1968).
Sharpe, M. R., *Satellites and Probes: The Development of Unmanned Spaceflight*. New York, 1970: Doubleday.
Thomas, P. G., "Crowding the Synchronous Orbit," *Space/Aeronautics*, Vol. 49, No. 4 (April 1968).
Wall, V., "Military Communications Satellites," *Astronautics & Aeronautics*, Vol. 6, No. 4 (April 1968).

CHAPTER 9

Medical Aspects of an Orbiting Research Laboratory, SP-86. Washington, D.C., 1966: National Aeronautics and Space Administration.
Problems and Uses of Outer Space. Pittsburgh, 1970: Carnegie-Mellon University.
Space Processing and Manufacturing, ME-69-1. Huntsville, Ala., 1969: Marshall Space Flight Center, National Aeronautics and Space Administration.
Space Research–Directions for the Future, Publication 1403. Washington, D.C., 1966: National Academy of Sciences and National Research Council.
STARLAB–Space Technology Applications and Research Laboratory, CR-61296. Auburn, Ala., 1969: Auburn University.
Bova, B., *The Uses of Space*. New York, 1965: Holt, Rinehart & Winston.
Gatts, J., *The Use of Space as a Research Tool*, AAS 67-122. Washington, D.C., 1967: American Astronautical Society.
Ruzic, N. P., *The Case for Going to the Moon*. New York, 1969: Putnam's.
Yaffee, M. L., "Space Factory Planned for 1970s," *Aviation Week & Space Technology*, Vol. 91, No. 19 (Nov. 10, 1969).

INDEX

(Page numbers in italics refer to illustrations.)